SpringerBriefs in Applied Sciences
and Technology

For further volumes:
http://www.springer.com/series/8884

Kazuo Matsuda · Yasuki Kansha
Chihiro Fushimi · Atsushi Tsutsumi
Akira Kishimoto

Advanced Energy Saving and its Applications in Industry

 Springer

Kazuo Matsuda
Chiyoda Corporation
4-6-2, Minatomirai, Nishi-ku
Yokohama 220-8765
Japan

Yasuki Kansha
Collaborative Research Center for
  Energy Engineering
Institute of Industrial Science
The University of Tokyo
4-6-1 Komaba, Meguro-ku
Tokyo 153-8505
Japan

Chihiro Fushimi
Department of Chemical Engineering
Tokyo University of Agriculture
  and Technology
2-24-16 Naka-cho, Koganei
Tokyo 184-8588
Japan

Atsushi Tsutsumi
Collaborative Research Center for
  Energy Engineering
Institute of Industrial Science
The University of Tokyo
4-6-1 Komaba, Meguro-ku
Tokyo 153-8505
Japan

Akira Kishimoto
Collaborative Research Center for
  Energy Engineering
Institute of Industrial Science
The University of Tokyo
4-6-1 Komaba, Meguro-ku
Tokyo 153-8505
Japan

ISSN 2191-530X   ISSN 2191-5318   (electronic)
ISBN 978-1-4471-4206-5   ISBN 978-1-4471-4207-2   (eBook)
DOI 10.1007/978-1-4471-4207-2
Springer London Heidelberg New York Dordrecht

Library of Congress Control Number: 2012938865

Printed on acid-free paper

Springer is part of Springer Science+Business Media (www.springer.com)

Preface

Refineries that occupy large sites in heavy chemical complexes and the petro-chemical industry in general, have for long years been utilizing and consuming a huge amount of fossil fuel as an energy source for operation. "Energy saving", in the way of reduction in the use of fossil fuel, has been under active consideration for many years as this leads to the strengthening of competitiveness by saving cost in operation. Moreover the industry in general fully recognizes that public opinion is less accepting of the combustion of fossil fuel that results in the generation and release of large amounts of the greenhouse gas CO_2 in to the atmosphere. Energy saving eventually leads to a significant reduction in the emission of greenhouse gases and is one of the most important measures that can be taken to mitigate the problem of global warming.

The sites in heavy chemical complexes have two systems in general; the process system and the utility system. The history of energy saving in the process system shows that its purpose was to improve and strengthen heat recovery. Pinch technology, a methodology of heat analysis based on thermodynamic principles, was developed in 1980s and applied for the study of heat recovery. Pinch technology is used to minimize the energy consumption of chemical processes by calculating thermodynamically feasible energy targets (or minimum energy consumption) and achieving them by optimizing heat recovery systems, energy supply methods and process operating conditions. Over the past 30 years, pinch technology has been applied on thousands of processes with a large amount of energy savings having been achieved.

However the general consensus in engineers was that almost all possible methods of heat recovery had already been investigated and that a new breakthrough was needed to be any further significant improvement in energy efficiency. That breakthrough was a new approach in the development of the process system, which would replace the conventional system.

The process system, in heavy chemical complexes, has a reaction section and a distillation section. In these sections, the amount of low-grade heat discarded as waste heat is about the same amount required to be supplied for the operation of the process. Conventionally heat recovery is maximized by using pinch

Fig. 1 From theory to applications about self-heat recuperation

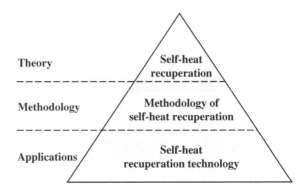

technology, which means that the amount of heat input is minimized and the waste heat is reduced. It should be noted that with this methodology the process operating condition is not changed but maintained as it is.

To achieve further energy saving in the process system, "self-heat recuperation" (SHR) was developed. (see Fig. 1) SHR is able to be applied for both the reaction section and the distillation section. SHR has a novel approach in that the process operating condition is changed by using a compressor, but the process core condition, such as a reactor inlet condition, is still maintained. Self-heat recuperation technology (SHRT) was applied to the reaction section of the naphtha HDS process and the benzene distillation section in the refinery and it was found that the advanced process with SHRT was able to reduce the energy consumption significantly for both the reaction section and the distillation section. SHRT is able to be applied not only to the process systems in the heavy chemical complexes but also to the other processes which require heating and cooling, such as drying and gas separation processes. It was found that SHRT was effective to reduce energy consumption considerably in such processes.

The utility system has also been targeted to become more efficient in energy saving. Major items of equipment, such as boilers and turbines, are considered to have reached the limit for further improvement in efficiency and therefore the utility system must be optimized. However, when looking at heavy chemical complexes, it appeared that a refinery would discharge low-grade heat as waste heat whereas, at same time, an adjacent petrochemical plant could make use of such heat. There could, therefore, be a large energy saving potential by utilizing the surplus heat across the sites. The total site approach, based on pinch technology, was applied to the heavy chemical complexes and it became apparent that there was a huge amount of energy saving potential through energy sharing among various sites in the complexes, despite the very high efficiency of the individual sites in the complex.

Tokyo, Japan

Kazuo Matsuda
Atsushi Tsutsumi

Contents

Part III   Utility System

Part I
Process System

Chapter 1
Energy Saving Technology

Abstract This chapter introduces the conventional and latest energy saving technologies for process systems, especially for use in oil refineries and petro-chemical plants. One of the most famous energy saving technologies for these processes is a well-known heat recovery technology that uses pinch technology. The hot and cold stream lines can be moved horizontally within the temperature limits in the temperature-heat diagram. Process systems are designed based on this graphical analysis. In contrast, in the latest energy saving technology termed self-heat recuperation technology, the hot stream line is shifted vertically by using the adiabatic compression of the hot stream in the temperature-heat diagram. Thus, the whole process heat can be recirculated into the process without any heat addition, leading to further energy saving in the process systems. In addition, process design methodology based on self-heat recuperation and the overall energy efficiency of the designed process are illustrated using simple thermal and distillation process examples.

Keywords Energy saving · Pinch technology · Self-heat recuperation technology · Process design · Process system · Exergy

1.1 Pinch Technology

Energy saving has, over several years, been attracting ever-increasing interest in many countries, as it would greatly assist in minimizing global warming caused mainly by the consumption of fossil fuels. Although many heat integration techniques for process energy saving have been applied to oil refineries since the 1970s' energy crisis, oil refineries, and petrochemical plants still consume large amounts of energy compared to the required values based on an exergy analysis. It

K. Matsuda et al., *Advanced Energy Saving and its Applications in Industry*,
SpringerBriefs in Applied Sciences and Technology,
DOI: 10.1007/978-1-4471-4207-2_1, © The Author(s) 2013

is commonly known that about 5 % of the amount of crude oil throughput in an oil refinery is used as fuel. In particular, about half of the total amount of fuel in an oil refinery is consumed in the crude oil distillation unit (CDU; atmospheric distillation). The CDU has one of largest fired heaters on the site, which consumes a huge amount of fuel oil and gas. In order to reduce the fuel consumption, CDUs had been equipped with a heat recovery system, which consists of approximately 10–20 heat exchangers, a so-called heat exchanger network system (HEN), leading to energy saving. However, it was very complicated to manage many hot and cold streams simultaneously in HEN. Thus, HEN was considered to be a prime candidate for implementation of a new methodology for optimization of its fuel consumption.

From 1980s, pinch technology has been applied to heavy chemical industries to determine suitable HENs based on a thermodynamic approach for energy saving. The concept of "target before design" was introduced by Linnhoff et al. (1982) using pinch technology for the design of individual processes. Pinch technology for HEN design was developed by Linnhoff and Hindmarsh (1983). Linnhoff and Ahmad (1990) and Ahmad et al. (1990) further evolved the methodologies to incorporate total cost targeting and block-decomposition based HEN synthesis. Later a HEN retrofit framework, based on the "process pinch" (Tjoe and Linnhoff 1986) and "network pinch" (Asante and Zhu 1996) concepts was established. Over time pinch technology has been applied to increasingly large and complex sites. To facilitate this, a variety of tools and techniques have been developed to enhance the methodology and simplify the analysis.

Figure 1.1 shows a conceptual temperature-heat diagram of heat integration systems by using conventional pinch technology (Eastop and Croft 1990; Kemp 2007). The lines which represent the cold and hot streams are plotted in the temperature-heat diagram. To handle multiple streams, the heat loads of all streams are added and a single composite of all hot streams and a single composite of all cold streams can be produced. They are called "Composite Curves." The cold stream means the process stream, in which the temperature is increasing, and the hot stream means the process stream, in which the temperature is decreasing. If $\Delta T_{\min}$ is negligible, the hot and cold stream lines overlap perfectly. In conventional pinch technology, these lines can be moved horizontally within the temperature limits until the nearest points (pinch points) are separated by the minimum temperature difference. From these plots, the lines can identify the region in which the heat is exchanged between the hot and the cold streams for heat recovery. Beyond the area of overlap, the curves identify the need for an additional heat source and sink. Pinch technology is based on a concept of energy cascading utilization, in which the heat energy is used from high grade heat to low grade heat and low grade waste heat is discarded from the process. Thus, pinch technology enables energy targets to be set without actual complete design and provides a consistent methodology for energy saving from the basic heat and material balance to the total site utility system.

Hence, pinch technology is well-known and has significant role of process design and optimization. In fact, it has now been applied to several heat exchanger networks.

Fig. 1.1 Temperature-heat diagram of heat integration

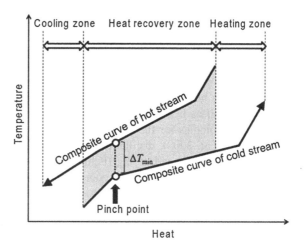

1.2 Self-Heat Recuperation Technology

The combustion of fossil fuels for heating produces a large amount of carbon dioxide (CO_2), which is the main contributor to the global greenhouse gas effect. Hence, the reduction of carbon dioxide (CO_2) emission and the reduction of energy consumption for heating has currently become a very important issue in the efforts to suppress global warming. Heat recovery technology such as pinch technology (Eastop and Croft 1990; Kemp 2007), which exchanges heat between the hot and cold streams in a process, has been applied to thermal processes to reduce energy consumption. A simple example of this technology is the application of a feed-effluent heat exchanger in thermal processes. In this heat exchanger, heat is exchanged between feed and effluent streams to recirculate the self heat of the stream (Seider et al. 2004). To exchange the heat, an additional heat source is required to provide temperature difference between two streams for heat exchange due to the second law of thermodynamics. These conventional heat recovery technologies are distinguished by cascading heat utilization, by which the required additional heat is provided by the exhausted heat from the other process or by the combustion of fuels. Although the net energy input seems to be reduced by using exhausted heat as the additional heat, the heat is also provided by the combustion of fossil fuels, leading to exergy destruction during energy conversion from chemical energy to heat (Som and Datta 2008).

Recently, attention has been paid to the analysis of process exergy and its irreversibility through consideration of the second law of thermodynamics. In many of these investigations, the calculation results of only exergy analysis and the possibility of the energy savings of some processes are only shown (Lampinen and Heillinen 1995; Chengqin et al. 2002; Aspelund et al. 2007; Grubbström 2007). From the process design point of view, a heat pump has been applied to thermal processes to reduce exergy destruction, in which the ambient heat or the process waste heat is generally pumped to heat the process stream by using working fluid compression (Fonyo and Benko 1996; Wu et al. 1998; Hou et al. 2007; Tarnawski 2009). Although it is well known that a heat pump can reduce energy consumption and exergy destruction in a process, the heat load and capacity of the process stream are often different from those of the pumped heat. Thus, a normal heat pump still possibly causes large exergy destruction during heating. As well as heat pumps for energy saving, vapor recompression in heat recovery technologies for the process has been applied to evaporation (Ettouney 2006; Nafey et al. 2008), distillation (Brousse et al. 1985; Annakou and Mizsey 1995; Haelssig 2008), and drying (Fehlau and Specht 2000), in which the vapor evaporated from the process is compressed to a higher pressure and then condensed, providing a heating effect. The condensation heat of the stream is recirculated as the vaporization heat in the process by using vapor recompression. However, many investigators have only focused on latent heat and few have paid attention to sensible heat. As a result, the total process heat cannot be recovered, indicating the potential for further energy savings in many cases.

Recently, an energy and exergy recuperative integrated gasification power generation system has been proposed and a design method for the system developed (Kuchonthara and Tsutsumi 2003; Kuchonthara et al. 2005; Kuchonthara and Tsutsumi 2006). Kansha et al. (2009) following on this concept developed self-heat recuperation technology (SHRT) based on exergy recuperation. The most important characteristic of this technology is that the entire process stream heat can be recirculated into a process without any heat addition, leading to marked reduction of exergy and energy savings for the process.

Self-heat recuperation (SHR; Kansha et al. 2009) facilitates recirculation, not only of the latent heat but also the sensible heat in a process, and helps to reduce the energy consumption of the process by using compressors and self-heat exchangers based on exergy recuperation. In this theory, (1) a process unit is divided on the basis of functions to balance the heating and cooling loads by performing enthalpy and exergy analysis and (2) the cooling load is recuperated by compressors and exchanged with the heating load. As a result, the heat of the process stream is perfectly circulated without heat addition, and thus the energy consumption for the process can be greatly reduced. Next, the process design methodology based on SHR is illustrated by using the simple thermal processes of the gas stream.

1.2.1 Process Design Methodology

Generally, separate process units each have functions such as heating, cooling, reaction, and separation. Figure 1.2 shows a thermal process without any heat recovery [I] in which a gas stream is heated from the standard temperature T_0 to a certain operating temperature T_1 of the following process by a heater ($1 \rightarrow 2$) and cooled to the standard temperature by cooler ($3 \rightarrow 4$). Note that, the following process is assumed not to affect the enthalpy of the process stream. The

Fig. 1.2 Simple thermal process [I] (for the gas stream): **a** flow diagram, **b** temperature-heat diagram

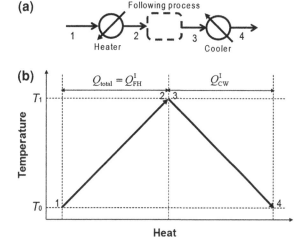

temperature-heat diagram of the thermal process [I] is shown in Fig. 1.2b. The total heating duty, Q_{total}, is provided by the heater. If the temperatures of the streams 2 and 3 are the same, the external heating load, Q_{FH}^I, is always equal to the external cooling load Q_{CW}^I.

In the case of a thermal process [II] using a feed-effluent heat exchanger as a representative example of conventional heat recovery with a minimum temperature difference ΔT_{min} (self-heat exchange thermal process), the heat of the effluent stream can be reused to preheat the feed stream up to T_2, reducing overall energy consumption, as shown in Fig. 1.3. In this process, the process stream is preheated from T_0 to T_2 with a heat exchanger (1→2) and heated with a heater from T_2 to T_1 (2→3). The effluent stream from the following process is cooled with the heat exchanger for self-heat exchange (4→5) and finally cooled to T_0 by a cooler (5→6). In Fig. 1.3b, the overlapping interval corresponds to the heat transfer duty from the hot effluent stream to the cold feed stream (the self-heat exchange load: Q_{HX}^{II}). Thus, heat recovery by means of self-heat exchange can reduce the external heating load (Q_{FH}^{II}). Note that Q_{CW}^{II} represents the external cooling load (5→6 in Fig. 1.3). Even with a self-heat exchange process, the heater is still required because the effluent stream should be heated to provide ΔT_{min} for heat exchange. Moreover, the heat generated by the heater should be dispersed in the cooler. The proposed thermal process of the gas streams for heat circulation based on SHR is shown in Fig. 1.4a. In this process, the feed stream is heated up with a heat exchanger (1→2) from a standard temperature T_0 to a set temperature T_1. The effluent stream from the following process is compressed with a compressor to recuperate the heat of the effluent stream (3→4) and the temperature at the exit of the compressor rises up to $T_{1'}$ because of the adiabatic compression. Stream 4 is cooled with the heat exchanger for self-heat exchange (4→5). The effluent stream is then decompressed with an expander to recover part of the work of the compressor. The effluent stream is finally cooled to T_0 with a cooler (6→7). Thus, SHR

Fig. 1.3 Thermal process
with a feed-effluent heat
exchanger [II] (for the gas
stream): **a** flow diagram,
b temperature-heat diagram

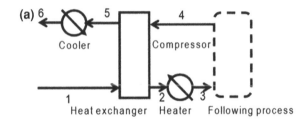

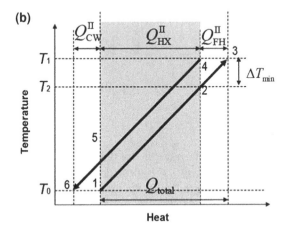

leads to perfect internal heat circulation. Note that the total heating duty, Q_{total}, is equal to the internal self-heat exchange load, Q_{HX}^{III}, without any external heating load as shown in Fig. 1.4b. In the case of ideal adiabatic compression and expansion, the input work provided to the compressor performs a heat pumping role in which the effluent temperature can achieve perfect internal heat circulation without exergy destruction. Therefore, SHR can dramatically reduce energy consumption. When designing the thermal process based on SHR, the enthalpy of inlet and outlet streams to the system must be equal. In this case, the enthalpy of stream 1 and 7, stream 2 and 3 must be equal. Otherwise, the energy saving efficiency of SHR is reduced.

As well as a gas stream case, a thermal process based on SHR in vapor/ liquid stream case can be developed as shown in Fig. 1.5. Figure 1.5a shows a thermal process for vapor/liquid streams with heat circulation based on SHR. In this process, the feed stream is heated with a heat exchanger (1→2) from a standard temperature, T_1, to a set temperature, T_2. The effluent stream from the subsequent process is pressurized by a compressor (3→4). The latent heat can then be exchanged between feed and effluent streams because the boiling temperature of the effluent stream is raised to T_b. by compression. Thus, the effluent stream is cooled through the heat exchanger for self-heat exchange (4→5) while recuperating its heat. The effluent stream is then depressurized by a valve (5→6) and finally cooled to T_1 with a cooler (6→7). This leads to perfect internal heat circulation based on SHR, similar to the above gas stream

Fig. 1.4 Thermal process based on self-heat recuperation [III] (for the gas stream): **a** flow diagram, **b** temperature-heat diagram

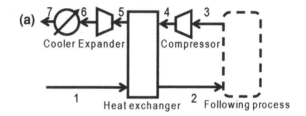

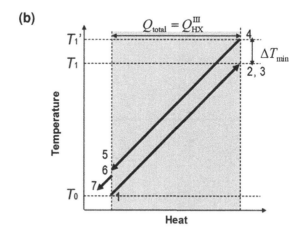

case. Note that, the total heating duty is equal to the internal self-heat exchange load without an external heating load, as shown in Fig. 1.5b. It is clear that the vapor and liquid sensible heat of the feed stream can be exchanged with the sensible heat of the corresponding effluent stream and the vaporization heat of the feed stream is exchanged with the condensation heat of the effluent stream. Similar to the thermal process for gas streams with heat circulation based on SHR, as mentioned above, the net energy required of this process is equal to the cooling duty in the cooler (6→7) and the exergy destruction occurs only during heat transfer in the heat exchanger. In SHR, the hot stream line of heat exchange is shifted vertically by using the adiabatic compression for the hot stream. The shaft work of the compression is required to circulate the internal heat in the process and exhausted with the process stream. If the heat capacity is independent from the pressure, the hot and cold stream lines are almost in parallel and are always separated by the minimum temperature difference. As a result, the energy required by the heat circulation module is reduced to 1/22–1/2 of the original by the feed-effluent heat exchange system in gas streams and/or vapor/liquid streams.

Fig. 1.5 Thermal process
based on self-heat
recuperation (for the vapor/
liquid phase change stream):
a flow diagram,
b temperature-heat diagram

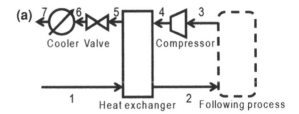

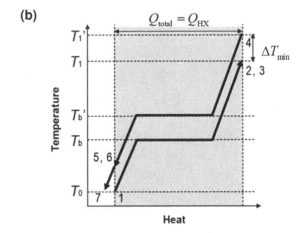

1.2.2 Design Methodology for Separation Process

Expanding the design methodology of the thermal process based on SHR to
separation processes (Kansha et al. 2010a, b), a system including not only the
separation process itself but also the preheating/cooling section, can be divided on
the basis of functions, namely the separation module and the heat circulation
module, in which the heating and cooling loads are balanced, as shown in Fig. 1.6.
To simplify the process for explanation, Fig. 1.6 shows a case that has one feed
and two effluents. In this figure, the enthalpy of inlet stream (feed) is equal to the
sum of the outlet streams (effluents) enthalpies in each module, giving an enthalpy
difference between inlet and outlet streams of zero. The cooling load in each
module is then recuperated by compressors and exchanged with the heating load
based on SHR. As a result, the heat of the process stream (self heat) is perfectly
circulated without the addition of heat in each module, resulting in perfect internal
heat circulation over the entire separation process.

   To understand this design methodology clearly, a binary distillation process is
used as an example of separation process for single feed dual effluents. By
following the above-mentioned design methodology, a distillation process can be
divided into two sections, namely the preheating and distillation sections, on the
basis of functions that balance the heating and cooling load by performing enthalpy
and exergy analysis, and both two sections are designed based on SHR. In the

Fig. 1.6 Conceptual figure of separation processes based on self-heat recuperation

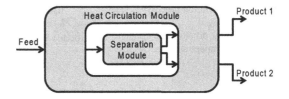

preheating section, one of the streams from the distillation section is a vapor stream and the stream to the distillation section has a vapor–liquid phase that balance the enthalpy of the feed streams and that of the effluent streams in the section. In balancing the enthalpy of the feed and effluent streams in the heat circulation module, the enthalpy of the streams in the distillation module is automatically balanced. Thus, the reboiler duty is equal to the condenser duty of the distillation column. Therefore, the vapor and liquid sensible heat of the feed streams can be exchanged with the sensible heat of the corresponding effluent streams and the vaporization heat can be exchanged with the condensation heat in each module. The vapor stream from the distillation section is compressed by a compressor. The preheating duty is supplied to the feed of the distillation system by the effluents of the distillation section via self-heat exchange. This preheating section is referred to as a heat circulation module for single feed and dual effluent streams. In the distillation section, the distillate is extracted as vapor from the distillation column and compressed by a compressor. The reboiler duty is supplied to the bottoms by the distillate via heat exchange. This distillation section is called a distillation module.

Figure 1.7a shows the structure of a distillation process based on SHR. This process consists of two standardized modules, the heat circulation module and the distillation module. Note that, the summation of the enthalpy of the feed streams and that of the effluent streams are equal in each module. The feed stream in this integrated process module is represented as stream 1. This stream is heated to its boiling point by the two streams recuperating heat of the distillate (12) and bottoms (13) by the heat exchanger (1→2). A distillation column separates the distillate (3) and bottoms (9) from stream 2. The distillate (3) is divided into two streams (4, 12). Stream 4 is compressed adiabatically by a compressor and cooled down by the heat exchanger (5→6). The pressure and temperature of stream 6 are adjusted by a valve and a cooler (6→7→8), and stream 8 is then fed into the distillation column as a reflux stream. Simultaneously, the bottom (9) is divided into two streams (10, 13). Stream 10 is heated by the heat exchanger and fed to the distillation column (10→11). Streams 12 and 13 are the effluent streams from the distillation module and return to the heat circulation module. In addition, the cooling duty of the cooler in the distillation module is equal to the compression work of the compressor in the distillation module because of the enthalpy balance in the distillation module. Furthermore, q value of column feed stream 2 (q = {heat needed to vaporize one mole of feed}/{molar latent heat of feed}) strictly depends on flow rate of distillate vapor and bottoms because of the enthalpy balance of distillation module.

Fig. 1.7 Distillation process based on self-heat recuperation: **a** process flow diagram, **b** temperature-heat diagram

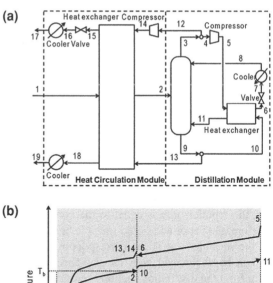

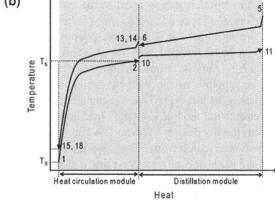

The effluent stream (12) from the distillation module is compressed adiabatically by a compressor (12→14). Streams 13 and 14 are successively cooled by heat exchangers. The pressure of stream 15 is adjusted to standard pressure by a valve (15→16), and the effluents are finally cooled to standard temperature by coolers (16→17, 18→19). The sum of the cooling duties of the coolers is equal to the compression work of the compressor in the heat circulation module. Streams 17 and 19 are the products.

Figure 1.7b shows the temperature and heat diagram for the SHR distillation process. In this figure, each number corresponds to the stream numbers in Fig. 1.7a, and T_s and T_b are the standard temperature and the boiling temperature of the feed stream, respectively. Both the sensible heat and the latent heat of the feed stream are subsequently exchanged with the sensible and latent heat of effluents in heat exchanger. The vaporization heat of the bottoms from the distillation column is exchanged with the condensation heat of the distillate from the distillation column in the distillation module. The heat of streams 4 and 12 are recuperated by the compressors and exchanged with the heat in the module. It can be seen that all the self heat is exchanged. As a result, the exergy loss of the heat exchangers can be minimized and the energy required by the distillation process is reduced to 1/6–1/8 of that required by the conventional heat exchanged distillation process.

1.2.3 Summary

Finally, by introducing the developed SHRT into several processes, process design methodology based on SHR can be summarized as followings:

1. A process unit is divided on the basis of function
2. Standardized module is modularized from divided process to balance the heating and cooling loads by performing enthalpy and exergy analysis
3. The appropriate heat pair (cooling load and heating load) is pointed out.
4. By following SHR, the self heat of process stream is recuperated and recirculated into the process without heat addition in each module.
5. The process is reconstructed from the integration of each module in which all of process self heat is recuperated and recirculated.

1.3 Conclusion

In this chapter, conventional energy saving technology (pinch technology) and the newly developed energy saving technology (self-heat recuperation technology) are introduced. The difference in these technologies and energy saving efficiencies is analyzed graphically. Moreover, the process design methodology of the energy saving process based on self-heat recuperation is summarized. In addition, the simulation results of the energy consumption are illustrated as compared with conventional counterparts.

References

Ahmad S, Linnhoff B, Smith R (1990) Targets and design for detailed capital cost models. Comput Chem Eng 14(7):751–767

Annakou O, Mizsey P (1995) Rigorous investigation of heat pump assisted distillation. Heat Recover Syst CHP 15(3):241–247

Asante NDK, Zhu XX (1996) An automated approach for heat exchanger retrofit featuring minimal topology modifications. Comput Chem Eng 20:s7–s12

Aspelund A, Bestad DO, Gundersen T (2007) An extended pinch analysis and design procedure utilizing pressure based exergy for subambient cooling. Appl Therm Eng 27:2633–2649

Brousse E, Claudel B, Jallut C (1985) Modeling and optimization of the steady state operation of vapor recompression distillation column. Chem Eng Sci 40(11):2073–2078

Chengqin R, Nianping L, Guangfa T (2002) Principle of exergy analysis in HVAC and evaluation of evaporative cooling schemes. Build Environ 37(11):1045–1055

Eastop TD, Croft DR (1990) Energy efficiency for engineers and technologists. Longman Scientific and Technical, London

Ettouney H (2006) Design of single-effect mechanical vapor compression. Desalination 190:1–5

Fehlau M, Specht E (2000) Optimization of vapor compression for cost savings in drying processes. Chem Eng Technol 23:901–908

Fonyo Z, Benko N (1996) Enhancement of process integration by heat pumping. Comuput Chem Eng 20:S85–S90

Grubbström RW (2007) An attempt to introduce dynamics into generalized exergy consideration. Appl Energy 84(7–8):701–718

Haelssig JB, Tremblay AY, Thibault J (2008) Technical and economical considerations for various recovery schemes in ethanol production by fermentation. Ind Eng Chem Res 47:6185–6191

Hou S, Li H, Zhang H (2007) Open air-vapor compression refrigeration system for air conditioning and hot water cooled by cool water. Energy Convers Manag 48:2255–2260

Kansha Y, Tsuru N, Sato K, Fushimi C, Tsutusmi A (2009) Self-heat recuperation technology for energy saving in chemical processes. Ind Eng Chem Res 48(16):7682–7686

Kansha Y, Tsuru N, Fushimi C, Shimogawara K. Tsutsumi A (2010a) An innovative modularity of heat circulation for fractional distillation. Chem Eng Sci 65(1): 330–334

Kansha Y, Tsuru N, Fushimi C, Tsutsumi A (2010b) Integrated process module for distillation procersses based on self-heat recuperation technology. J Chem Eng Jpn 43(6):502–507

Kemp IC (2007) Pinch analysis and process integration A user guide on process integration for the efficient use of energy 2nd Edn. Elsevier, Oxford

Kuchonthara P, Tsutsumi A (2003) Energy-recuperative biomass integrated gasification power generation system. J Chem Eng Jpn 36(7):846–851

Kuchonthara P, Bhattacharya S, Tsutsumi A (2005) Combination of thermochemical recuperative coal gasification cycle and fuel cell for power generation. Fuel 84(7–8):1019–1021

Kuchonthara P, Tsutsumi A (2006) Energy-recuperative coal-integrated gasification/gas turbine power generation system. J Chem Eng Jpn 39(5):545–552

Lampinen MJ, Heillinen MA (1995) Exergy analysis for stationary flow systems with several heat exchange temperatures. Int J Energy Res 19(5):407–418

Linnhoff B, Townsend DW, Boland D, Hewitt GF, Thomas BEA, Guy AR, Marsland RH (1982) A user guide on process integration for the efficient use of energy, 1st edn. Inst Chem Eng, Rugby

Linnhoff B, Hindmarsh E (1983) The pinch design method of heat exchanger networks. Chem Eng Sci 38(5):745–763

Linnhoff B, Ahmad S (1990) Cost optimum heat exchanger networks-1. minimum energy and capital using simple models for capital cost. Comput Chem Eng 14(7):729–750S

Nafey AS, Fath HES, Mabrouk AA (2008) Thermoeconomic design of a multi-effect evaporation mechanical vapor compression (MEEMVC) desalination process. Desalination 230:1–15

Seider WD, Seader JD, Lewin DR (2004) Product and process design principles synthesis, analysis, and evaluation 2nd Edn. Wiley, New York

Som SK, Datta A (2008) Thermodynamic irreversibilities and exergy balance in combustion processes. progress in energy combustion. science 34(3):351–376

Tarnawski VR, Leong WH, Momose T, Hamada Y (2009) Analysis of ground source heat pumps with horizontal ground heat exchangers for northern Japan. Renew Energy 34:127–134

Wu C, Chen L, Sun F (1998) Optimization of steady flow heat pumps. Energy Convers Manag 39(5/6):445–453

Part II
Application of Self-Heat Recuperation Technology

Chapter 2
Reaction Section

Abstract The naphtha hydrodesulfurization (HDS) process in the refinery has a reaction section which is a heating and cooling thermal process consisting of a feed-effluent heat exchanger and a fired heater. Energy savings are fundamentally made as a result of the maximized heat recovery in the heat exchanger and the reduced heat duty of the fired heater. To achieve further energy saving in the process, "self-heat recuperation technology" (SHRT) was adopted by a naphtha HDS process with the capacity of 18,000 barrel per stream day (BPSD). A compressor was introduced in this technology. The suction side of the compressor needed lower pressure so that the feed stream could evaporate more easily. The discharged side of the compressor satisfied the operating conditions of both pressure and temperature at the inlet of the reactor. And the reactor effluent stream was able to be used completely to preheat and vaporize the feed stream. All the heat in the process stream was recirculated without the necessity of a fired heater. It was confirmed that despite there being no more energy saving potential in the conventional process that makes use of a fired heater, the advanced process based on SHR can reduce the energy consumption significantly by using the recuperated heat of the feed stream.

Keywords Self-heat recuperation technology · Reactor · Naphtha · Hydro desulfurization · Compressor · Minimum temperature difference

2.1 Introduction

The naphtha hydrodesulfurization (HDS) process, a well-known process adopted by refineries, uses a fired heater in the reaction section to heat the process streams. The reaction section of such process is a heating and cooling thermal process

K. Matsuda et al., *Advanced Energy Saving and its Applications in Industry*,
SpringerBriefs in Applied Sciences and Technology,
DOI: 10.1007/978-1-4471-4207-2_2, © The Author(s) 2013

consisting of a feed-effluent heat exchanger and a fired heater. The fired heater consumes not only a large amount of energy but also a large amount of exergy during combustion due to the large temperature difference in the fired heater between fuel combustion (over 800 °C) and process condition (around 300 °C) for the reactor.

Energy savings are fundamentally made as a result of the maximized heat recovery in the heat exchanger and the reduced heat duty of the fired heater. In order to reduce energy consumption in the fired heater, pinch technology has been widely applied for heat recovery to save energy in a plant or a complex of plants. To achieve further energy saving in the process, "self-heat recuperation technology" (SHRT) was adopted (Matsuda et al. 2010).

2.2 Process Flow Description

Figure 2.1 shows the naphtha HDS process (conventional case). Liquid naphtha is pumped and mixed with compressed hydrogen and then the mixed stream of naphtha and hydrogen gas, which has a vapor/liquid mixed phase, is heated in a heat exchanger. The stream is further heated by a fired heater to the required condition (300 °C) for the reactor. Because of the temperature difference between the process requirement (300 °C) and the inside of the fired heater (>800 °C), fuel combustion causes a large exergy loss. The reactor effluent stream is cooled down by the heat exchanger and the cooling water cooler.

2.3 Applying Self-Heat Recuperation Technology

Figure 2.2 shows the application of SHRT (proposed case) to naphtha HDS process. A reactor charge compressor is installed and the heat exchanger is revamped to increase its surface area. The liquid naphtha is pumped and mixed with hydrogen gas without the necessity of a hydrogen compressor, and then the mixed stream of naphtha and hydrogen gas is heated and vaporized totally in the heat exchanger by the effluent heat. Finally, the feed stream is compressed by the reactor charge compressor to satisfy the operating conditions of both pressure and temperature for the reactor. The heat of the reactor effluent stream completely preheats and vaporizes the feed stream. As a result, the self-heat of the process streams can be recirculated in the process without requiring any additional heat source, leading to the reduction in the aforementioned exergy loss.

Fig. 2.1 Conventional case
of naphtha HDS process

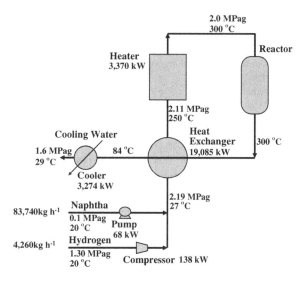

Fig. 2.2 Proposed case

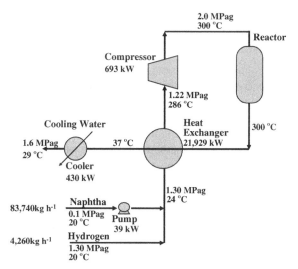

2.4 Prospect of Applying Self-Heat Recuperation Technology

Tsuru et al. 2008 and Kansha et al. 2009 reported that SHRT was used for heating
and cooling in thermal processes by pressure change and it could achieve perfect
internal heat circulation in a feed-effluent heat exchanger as shown in the
temperature-heat diagram of the proposed case from Sect. 1.2. The reactor charge
compressor is used in the naphtha HDS process as shown in Fig. 2.2. By
compressing the stream, the stream pressure and its temperature are both increased
simultaneously. The feed stream to the heat exchanger needs to be set at a lower

Table 2.1 Study basis

Case	Conventional Case-1		Case-2 [ultimate]	Proposed
Minimum temperature difference in heat exchanger, K	16.2	11.5	0.01	11.5
Heat exchanged, kW	19,085	19,806	21,658	21,929
	(−)	(+4 %)	(+13 %)	(+15 %)

pressure than that in the conventional naphtha HDS process (conventional case), which then allows the reactor inlet pressure in the proposed case to remain the same as in the conventional case. As explained in Sect. 1.2, the boiling temperature is shifted from T_b to $T_{b'}$ by the pressure change of the feed stream, which allows the latent heat to be exchanged between the feed and effluent streams. All the heat of the process stream is recirculated in the process without the necessity of a fired heater, which means that the heat of the reactor effluent stream can be used completely to preheat and vaporize the feed stream, resulting in a drastic reduction in the energy consumption of the process.

2.5 Mass and Heat Balance Calculation

The energy required for the conventional and the proposed cases of the naphtha HDS process in a refinery was calculated by using a commercially available simulator, PRO/II™ version 8.1 (Invensys). The Soave–Redlich–Kwong equation of state was applied. Figures 2.1 and 2.2 show the results of the case studies. The throughput of the naphtha HDS process was 18,000 barrel per stream day (BPSD) with the inlet condition of the reactor set to 2.0 MPag and 300 °C. Table 2.1 shows the conditions of the minimum temperature difference (ΔT_{min}) in the case studies. In the conventional case (current operating condition in a commercial plant), ΔT_{min} was 16.2 K. Another simulation was then run, for comparison purposes, under ΔT_{min} (11.5 K) to establish the heat and pressure balance around the compressor for proposed case and conventional process (Case-1). In order to confirm the thermodynamic limit in the conventional process, Case-2 (0.01 K) was prepared as the ultimate condition requiring an infinitely large surface area.

2.5.1 Conventional Case

As shown in Fig. 2.1, the naphtha pump and hydrogen compressor in the conventional case raised the pressure of the streams to 2.19 MPag, so as to meet the required pressure at the inlet of the reactor. After heat recovery through the heat exchanger, the stream was 2.11 MPag and 250 °C at the inlet of fired heater, which was then heated to 300 °C in the fired heater. The heat duty (3,274 kW) of

Table 2.2   Study results

Case	Conventional	Proposed	
Minimum temperature difference in heat exchanger, K	16.2	11.5	
1. Exergy input, kW			
1) Fired heater	3,370	–	
2) Naphtha pump	68	39	
3) Hydorgen compressor	138	–	
4) Reactor charge compressor	–	693	
Total	3,576 (100 %)	732 (20 %)	
2. Energy input, kW			Efficiency
1) Fired heater	3,965	–	1) 85 %
2) Naphtha pump	310 a	172 b	2) a: 60 %, b: 62 %
3) Hydorgen compressor	454	–	3) 83 %
4) Reactor charge compressor	–	2,014	4) 94 %
Total	4,729 (100 %)	2,186 (46 %)	(2)–(4) Power generation efficiency: 36.6 %

the cooling water cooler was almost the same as the heat duty (3,370 kW) of a fired heater, which meant that this process consumed a large amount of high-grade heat in the fired heater and exhausted a large amount of low-grade heat in the cooling water cooler.

Table 2.1 indicates that the heat duty in the heat exchanger increased, as Case-1 and Case-2 were 4 and 13 % larger respectively than that in the conventional case. In Cases 1 and 2, ΔT_{min} decreased, meaning that surface area of the heat exchanger increased compared with conventional case.

Table 2.2 shows the exergy and energy input to the naphtha HDS process (conventional case). The sum of exergy input in conventional case (a fired heater, a naphtha pump and a hydrogen compressor) was 3,576 kW and the sum of energy input was 4,729 kW.

2.5.2 Proposed Case

In the proposed case (Fig. 2.2), the installation of a compressor resulted in a decrease in the pressure of the feed stream at the inlet of the heat exchanger (1.30 MPag) compared with the conventional case (2.19 MPag), which reduced the naphtha pump power from 68 to 39 kW. Because hydrogen gas at 1.3 MPag was supplied from an outside process, it could be fed directly in the proposed case without the requirement for a hydrogen compressor, resulting in further energy saving. The power of the compressor was 693 kW, which was only 20 % of the heat duty of the fired heater in conventional case (3,370 kW). The heat duty (430 kW) of the cooling water cooler was almost the same as the power of the compressor.

As can be seen in Table 2.1, the heat exchange duty in the proposed case, which was 11 % more than that of Case-1, could be increased by 15 % as compared with the conventional case.

In the proposed case in Table 2.2, the sum of exergy input (a naphtha pump and a reactor charge compressor) was 732 kW, which was only 20 % of the conventional case. Taking equipment efficiency and power generation efficiency into account (36.6 % in the Japanese energy saving law), the sum of energy (enthalpy) input was 4,729 kW in the conventional case and 2,186 kW in the proposed case, and the energy consumption in the proposed case was 46 % of that in the conventional case. The amount of energy saving was calculated to be 1,900 kL y^{-1} (kiloliter per year) by annual crude oil conversion. This is equivalent to 7.4×10^4 GJ y^{-1} (gigajoule per year).

2.6 Discussion

The design strategy of a heat pump was recently introduced, with process integration under optimal matching (Fonyo and Benko 1996; Wu et al. 1998; Smith et al. 2005). Energy and exergy analysis of the heat pump was applied by Ceylan et al. 2007. Pavlas et al. 2010 designed the heat pump system with the aid of Grand Composite Curve. The heat pump was applied to several thermal processes; the liquefaction process of natural gas (Aspelund et al. 2007), the open air–vapor compression refrigeration system (Hou et al. 2007), multi-function heat pump system (Gong et al. 2008), the ground source heat pump system (Tarnawski et al. 2009), and the refrigeration process using moist air and water (Hou and Zhang 2009). The ambient heat or the process waste heat is generally pumped by the heat pump to heat/cool the process stream by the working fluids (steam, CO_2, etc.). Because the temperature difference between the heat source and the feed stream is much larger than that in the heat exchanger in proposed process, the heat pump system requires more power to improve the quality of the heat from the heat source to the heater condition than the power required in the proposed process based on SHR.

As one of the heat recovery technologies, the vapor recompression technology has been applied to evaporation (Ettouney 2006; Nafey and Fath 2008), distillation (Brousse et al. 1985; Annakou and Mizsey 1995; Haelssig et al. 2008), and drying (Fehlau and Specht 2000). This technology compresses the vapor to a higher pressure and then the pressurized vapor provides a heating effect when condensing. However, such technology did not utilize the condensation heat efficiently in the heat exchanger. In contrast, SHRT can utilize the condensation heat thoroughly by changing the shape of the temperature-heat line in the heat exchanger because the installation of a compressor results in a change in the pressure balance. The feed stream is completely preheated and vaporized by the heat of the reactor effluent stream, which means that all the heat of the process streams was recirculated in the process without needing to use a fired heater. Consequently, SHRT could reduce

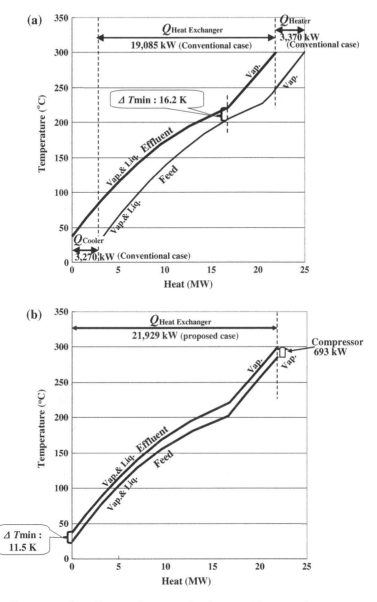

Fig. 2.3 Temperature-heat diagram of **a** conventional case and **b** proposed case

the exergy loss in a fired heater and in a cooling water cooler. It should be noted that the required heat input in the proposed case was obtained not by firing fuel but by working power.

Figure 2.3 shows the temperature-heat diagram for feed and effluent streams in the conventional and proposed cases, respectively. The feed stream and the effluent stream are both represented in Fig. 2.3a. The feed stream, which is a mixed stream of naphtha and hydrogen gas, has a pinch point (ΔT_{min} = 16.2 K) against the effluent stream in the conventional case. By following pinch technology, the cold composite curve is moved horizontally within the temperature limits to pinch point in order to increase the heat recovery. By increasing the surface area in the conventional case, the feed stream could be moved closer, horizontally. This would be Case-1 whereby ΔT_{min} became 11.5 K, resulting in an increase of 4 % in the heat recovery as shown in Table 2.1.

In Fig. 2.3b, the installation of the reactor charge compressor in the proposed case can reduce the pressure of the feed stream at the heat exchanger compared with the conventional case. By decreasing its pressure, the feed stream was moved and finally approached the effluent stream much more closely. As a result, the heat recovery was vastly increased to 21,929 kW, 15 % more than the conventional case, where ΔT_{min} was 11.5 K. It should be noted that the amount of heat recovery (21,929 kW) in the proposed case was still more than 21,658 kW of the ultimate condition in Case 2 as shown in Table 2.1, which indicates that the SHRT is effective to increase energy efficiency.

2.7 Conclusion

In the naphtha HDS process (conventional case) with a fired heater, the heat energy was heretofore cascaded from high-grade heat to low-grade heat and then low-grade waste heat was discarded from the process. The advanced process using SHRT (proposed case) was developed to utilize the heat of the stream itself by effectively introducing a reactor charge compressor. The process heat energy itself was recuperated and utilized based on exergy recuperation, which ensured that the necessary heat input was obtained not by firing at the fired heater but by working power from the reactor charge compressor. As the heat exchange duty could be increased by 15 %, the energy and the exergy inputs in proposed case were significantly reduced (46 and 20 %) as compared with the conventional case. The cooling water cooler duty could also be decreased considerably from 3,274 to 430 kW.

References

Annakou O, Mizsey P (1995) Rigorous investigation of heat pump assisted distillation. Heat Recovery Syst CHP 15(3):241–247
Aspelund A, Berstad DO, Gundersen T (2007) An extended pinch analysis and design procedure utilizing pressure based exergy for subambient cooling. Appl Therm Eng 27:2633–2649
Brousse E, Claudel B, Jallut C (1985) Modeling and optimization of the steady state operation of vapor recompression distillation column. Chem Eng Sci 40(11):2073–2078
Ceylan I, Aktas M, Dogan H (2007) Energy and exergy analysis of timber dryer assisted heat pump. Appl Therm Eng 27:216–222

Ettouney H (2006) Design of single-effect mechanical vapor compression. Desalination 190:1–5

Fehlau M, Specht E (2000) Optimization of vapor compression for cost savings in drying processes. Chem Eng Technol 23:901–908

Fonyo Z, Benko N (1996) Enhancement of process integration by heat pumping. Comput Chem Eng 20:S85–S90

Gong G, Zeng W, Wang L, Wu C (2008) A new heat recovery technique for air-conditioning/ heat-pump system. Appl Therm Eng 28:2360–2370

Haelssig JB, Tremblay AY, Thibault J (2008) Technical and economical considerations for various recovery schemes in ethanol production by fermentation. Ind Eng Chem Resour 47:6185–6191

Hou S, Li H, Zhang H (2007) Open air-vapor compression refrigeration system for air conditioning and hot water cooled by cool water. Energy Convers Manag 48:2255–2260

Hou S, Zhang H (2009) An open reversed Brayton cycle with regeneration using moist air for deep freeze cooled by circulation water. Int J Therm Sci 48:218–223

Kansha Y, Tsuru N, Sato K, Fushimi C, Tsutsumi A (2009) Self-heat recuperation technology for energy saving in chemical processes. Ind Eng Chem Res 48:7682–7686

Matsuda K, Kawazuishi K, Hirochi Y, Sato R, Kansha Y, Fushimi C, Shikatani Y, Kunikiyo H, Tsutsumi A (2010) Advanced energy saving in the reaction section of the hydro-desulfurization process with self-heat recuperation technology. Appl Therm Eng 30:2300–2305

Nafey AS, Fath HES, Mabrouk AA (2008) Thermoeconomic design of a multi-effect evaporation mechanical vapor compression (MEE-MVC) desalination process. Desalination 230:1–15

Pavlas M, Stehlík P, Oral J, Klemes J, Kim JK, Firth B (2008) Heat-integrated heat pumping for improved energy management case study: biomass gasification in wood processing plant, 18th International Congress of Chemical and Process Engineering (CHISA 2008), 24th-28th August, Czech Republic, Prague

Smith JM, Van Ness HC, Abbott MM (2005) Introduction to chemical engineering thermody-namics, 7th edn. McGaw-Hill, New York

Tarnawski VR, Leong WH, Momose T, Hamada Y (2009) Analysis of ground source heat pumps with horizontal ground heat exchangers for northern Japan. Renew Energy 34:127–134

Tsuru N, Sato N, Kansha Y, Fushimi C, Shimogawara K, Tsutsumi A (2008) Self-heat recuperation technology for sustainable chemical process. In: Proceeding of 20th International Symposium on Chemical Reaction Engineering. pp 506–507

Wu C, Chen L, Sun F (1998) Optimization of steady flow heat pumps. Energy Convers Manag 39(5/6):445–453

Chapter 3
Distillation Section

Abstract There are several types of distillation processes, the most commonly used of which is the distillation process for a binary system. In this process, heat is supplied at the feed heater and reboiler, and the overhead stream is cooled at a condenser. Almost all of the supplied heat at the reboiler in the conventional distillation process is discarded in the overhead condenser. Conventional energy savings in the distillation processes were fundamentally attained as a result of heat recovery duty in the feed heater being maximized by using the heat of the bottom stream, which enabled the utility (steam or hot oil) rate to the feed heater to be reduced. "Self-heat recuperation technology" (SHRT) was adopted to achieve further energy saving in the distillation process, whereby two compressors are installed in the overhead vapor line, consisting of the reflux and the overhead product streams. A compressor (compressor-1) treats the reflux stream and the other compressor (compressor-2) treats the overhead stream. The reboiler duty is supplied by the recuperated heat of the discharged stream from compressor-1 and the feed heater duty is supplied by that from compressor-2, by adiabatic compressions. It was able to be determined that the advanced process based on self-heat recuperation (SHR) could reduce the energy consumption significantly by using the recuperated heat of the overhead vapor.

Keywords Self-heat recuperation technology · Distillation · Binary system · Benzene · Compressor · Overhead · Reboiler

3.1 Introduction

The distillation process for binary system consists of a feed stream and two product streams, i.e., an overhead product stream and a bottom product stream. Heat is supplied at a feed heater in the feed stream and a reboiler at the bottom of

K. Matsuda et al., *Advanced Energy Saving and its Applications in Industry*,
SpringerBriefs in Applied Sciences and Technology,
DOI: 10.1007/978-1-4471-4207-2_3, © The Author(s) 2013

the distillation column, and the overhead stream is cooled at a condenser. The feed stream is heated by the heat of the bottom product stream and then is further heated by steam or hot oil. The reboiler is also heated by steam or hot oil and sometimes by a heating furnace (reboiler furnace). Almost all of the supplied heat at the reboiler in the distillation process is discarded in the overhead condenser.

As a conventional energy saving method for a distillation process in industry, the heat recovery duty to the feed heater was maximized, but little consideration was given to the heat recovery duty of the reboiler.

For further energy saving, heat integration methods were developed for distillation columns, such as vapor recompression (VRC) distillation columns (Annakou and Mizsey 1995; Enweremadu et al. 2009; Jogwar and Daoutidis 2009) and heat integrated distillation columns (HIDiC: Campbell et al. 2008, Wang et al. 2009, Suphanit 2010). A design strategy was introduced of a heat pump with process integration under optimal matching (Fonyo and Benko 1996; Wu et al. 1998; Smith et al. 2005). Energy and exergy analysis of the heat pump was applied by Ceylan et al. (2007). Pavlas et al. (2010) designed the heat pump system with the aid of Grand Composite Curve (Dhole and Linnhoff 1992; Raissi 1994; Klemes et al. 1997; Perry et al. 2008). The heat pump was applied to several thermal processes; the liquefaction process of natural gas (Aspelund et al. 2007), the open air-vapor compression refrigeration system (Hou et al. 2007), multifunction heat pump system (Gong et al. 2008), the ground source heat pump system (Tarnawski et al. 2009), and the refrigeration process using moist air and water (Hou and Zhang 2009). The ambient heat or process waste heat is generally pumped by a heat pump to heat/cool the process stream by working fluids (steam, CO_2, etc.).

Maximizing the heat recovery duty in the feed heater by using the heat of the bottom stream enabled the utility (steam or hot oil) rate to the feed heater to be reduced. "Self-heat recuperation technology" (SHRT) was adopted (Matsuda et al. 2011) to achieve further energy saving in the distillation process.

In this technology, two compressors are installed in the overhead vapor line, which consists of the reflux and the overhead product streams. A compressor (compressor-1) treats the reflux stream and the other compressor (compressor-2) treats the overhead stream. The reboiler duty is supplied by the recuperated heat of the discharged stream from compressor-1 and the feed heater duty is supplied by that from compressor-2, by adiabatic compression.

3.2 Process Flow Description

A conventional distillation process flow diagram is shown in Fig. 3.1. The heat to the distillation column is supplied by a feed heater (E1) and reboiler (E3) using the utility steam or hot oil. The overhead vapor is subcooled by an overhead condenser (E2) and the condensed liquid is routed into the overhead drum. One of the liquid streams from the overhead drum is the reflux stream, which is sent back to the top tray of the distillation tower, and the other is the overhead product stream.

The overhead product stream and the product stream from the bottom of the distillation tower are further cooled by separate rundown coolers (E4, E5).

3.3 Study Basis

The operating condition (conventional case) of the distillation process in this study was determined to provide the benchmark process for comparison. The distillation process was divided into the two envelopes (an inner one and outer one) as shown in Fig. 3.1. The enthalpy of stream A (H_A) in the inner envelope was equal to the sum of the enthalpy of streams B (H_B) and C (H_C) as shown in Eq. (3.1). The overhead cooling duty (E2) was adjusted to be the same as the reboiler (E3) duty. Here, the overhead vapor was cooled to bubble point condition. The enthalpy of stream 1 (H_1) in the outer envelope was equal to the sum of the enthalpy of streams 2 (H_2) and 3 (H_3) as shown in Eq. (3.2). Streams 1–3 were set in the standard condition, such as 25 °C and 0.10 MPag.

$$H_A = H_B + H_C \qquad (3.1)$$

$$H_1 = H_2 + H_3 \qquad (3.2)$$

3.4 Applying Self-Heat Recuperation Technology

The SHRT was applied to a distillation process to provide the new process flow (proposed case). In such case as shown in Fig. 3.2, two compressors were installed in the overhead vapor line. The overhead vapor consisted of the reflux stream and the overhead product stream. It should be noted that compressor-1 (C1) treated the reflux stream and compressor-2 (C2) treated the overhead product stream. The temperatures of the discharged streams were able to be increased by adiabatic compression. The reboiler (E3) duty was supplied by the heat of the compressed stream discharged from compressor-1 (C1). The heat input after reboiler E3 was adjusted by cooler E4 in the return line to the distillation column and the feed heater (E2) duty was supplied by the heat of the compressed stream discharged from compressor-2 (C2). The heat of the feed stream at another feed heater (E1) was exchanged with the heat of the bottom product stream. Finally, the overhead product stream and bottom product stream were cooled to the standard condition (25 °C and 0.10 MPag) by coolers (E5, E6).

3.5 Prospect of Applying Self-Heat Recuperation Technology

In the VRC and HIDiC methods for energy saving for the distillation process, latent heat (condensation heat) is recovered by vapor compression thereby exploiting energy cascading to reduce the heating load. However, in both VRC and

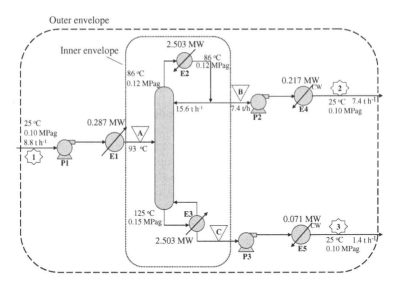

Fig. 3.1 Distillation process (conventional case)

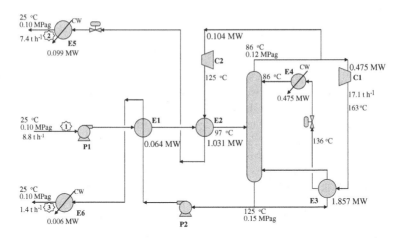

Fig. 3.2 Proposed case

HIDiC, only the heat recovery duty to the reboiler in the distillation column is considered, but the heat during preheating is not or is less recognized.

On the contrary, Kansha et al. (2009, 2010a, b) have recently developed a novel SHRT that utilizes not only latent heat in the process but also sensible heat by using compressors and self-heat exchangers based on exergy recuperation. It means that a reboiler utilizes latent heat and a feed heater utilizes both latent and sensible heats based on SHR. As a principle of SHR, (i) a process unit is divided on the basis of functions to balance the heating and cooling load by performing enthalpy and exergy analysis, and (ii) the cooling load is recuperated by

Table 3.1 Component balances

Stream components	Feed (wt %)	Overhead (wt %)	Bottom (wt %)
Benzene	85.02	99.94	4.07
Toluene	12.83	0.00	82.38
C_8H_{18}	0.20	0.06	0.95
Et-Benzene	0.14	0.00	0.89
Xylene	1.81	0.00	11.71

compressors and exchanged with the heating load. The overhead vapor is pressurized by compressors and its temperature is increased by adiabatic compression. The duties of reboiler and feed heater are supplied by the heat of the discharged streams from the compressors. As a result, the heat of the process stream is circulated without the need for any additional heat and, thus, the energy consumption of a process can be greatly reduced.

3.6 Mass and Heat Balance Calculation

The benzene distillation process was studied as an industrial application. The feedstock, under operating conditions, was a mixture of benzene, toluene, and other constituents. The feed rates of the feedstock, the overhead product and the bottom product were 8.8, 7.4 and 1.4 t/h, respectively. The component balance for these three streams was specified as shown in Table 3.1. The commercial process simulator, PRO/II^{TM} (Invensys, Ver. 8.1), and the Soave–Redlich–Kwong equation of state (SRK) were used. The material and energy balances in both conventional and proposed cases were calculated.

3.6.1 Conventional Case

As shown in Fig. 3.1, the feed heater (E1) required 0.287 MW and the reboiler required 2.503 MW. Here, the overhead vapor was cooled to bubble point condition of 86 °C and routed into the overhead drum. Table 3.2 shows the exergy and energy input (as fuel) for the conventional case. The sum of exergy input for the conventional case with a feed heater and a reboiler was 2.790 MW and the sum of energy input was 3.100 MW. The feed heater and reboiler were heated by utility, such as steam or hot oil, which was supplied by a boiler or a furnace heater. In this calculation, the efficiency of the equipment was taken into account as 90 %. In the conventional case, several pumps were powered by electricity, but the power consumption was negligible.

Table 3.2 Study results

Case	Conventional	Proposed (two compressors)	Economical (single compressor)	Remarks
1. Exergy input, MW				
(1) Feed heater	(E1) 0.287	–	–	
(2) Reboiler	(E3) 2.503	–	–	
(3) Compressor-1	–	0.475	0.682	
(4) Compressor-2	–	0.104	–	
Total	2.790 (100 %)	0.579 (20.7 %)	0.682 (24.4 %)	
2. Energy input, MW				
(1) Feed heater	(E1) 0.319	–	–	Equipment efficiency 90 %
(2) Reboiler	(E3) 2.781	–	–	
(3) Compressor-1	–	1.298	1.863	Compressor efficiency 65 %
(4) Compressor-2	–	0.284	–	Power efficiency 36.6 %
Total	3.100 (100 %)	1.582 (51.0 %)	1.863 (60.1%)	

3.6.2 Proposed Case

In the proposed case as shown in Fig. 3.2, two compressors were installed in the overhead vapor line. Compressor-1 required 0.475 MW and compressor-2 required 0.104 MW, when the compressor efficiency was 65 %. Table 3.2 shows the exergy and energy input for proposed case. The sum of exergy input for proposed case (compressors-1 and -2) was 0.579 MW, which equated to 20.7 % of the conventional case, and the sum of energy input was 1.582 MW, equivalent to 51.0 % of the conventional case. Note that, the primary energy input was calculated based on a Japanese power generation efficiency of 36.6 %.

3.7 Discussion

The VRC technology, a well-known heat recovery technology, was applied to evaporation (Ettouney 2006; Nafey et al. 2008), distillation (Brousse et al. 1985; Annakou and Mizsey 1995; Haelssig et al. 2008), and drying (Fehlau and Specht 2000). Hirata (2009) recently evaluated the technology, in which the vapor is compressed to a higher pressure and then the pressurized vapor provides a heating effect while condensing. However, such technology utilizes only the latent heat but not the sensible heat.

In contrast, SHRT utilizes not only latent heat but also the sensible heat in the process by using a compressor. In SHRT, the heat of the streams is recuperated by using compressors and the recuperated heat of the pressurized stream is supplied to the cold stream by heat exchange.

In the distillation process for the heavy chemical industrial field, it can be said that the two types of processes based on SHR were integrated. One is SHR distillation in which the partial heat of the overhead vapor (the reflux stream) is recuperated and the recuperated heat is supplied to the reboiler. The other is the heat circulation system for feed heating, which leads to the remaining partial heat of the overhead vapor (the overhead product stream) being recuperated and being supplied to the feed heater. As a result, all the heat is recirculated in the process without the need for an external heat source and only electric power for the compressors is required. In the heat recovery network for the feed stream in the case study of Kansha et al. (2010b), the feed stream was divided into two parallel streams, which were exchanged with the overhead and the bottom product streams to maximize the heat recovery duty. However, taking into account the actual component balance, the practical and compact heat recovery network for industrial application, as shown in Fig. 3.2, was prepared by sacrificing some energy reduction because it may be possible, but not practical, to improve the energy efficiency by optimizing the heat recovery. By applying SHRT, the energy consumption of the proposed case could be substantially decreased to 51.0 % of the conventional case, as shown in Table 3.2.

From the point of view of economical process design (Asprion et al. 2011; Van der Ham and Kjelstrup 2010), the cost of installing compressors increases the total project cost of the proposed case appreciably as compared to the conventional case. Hence, a further case study was conducted from the viewpoint of cost saving when a compressor was introduced. An economical case was developed and investigated when designing the process flow in the proposed case. In the economical case, instead of the two compressors being installed in the proposed case, only one compressor was installed as shown in Fig. 3.3. The compressor treated all overhead vapor and the compressed vapor was divided into the reboiler and the feed heater. Although the economical case could not reduce energy consumption to the minimum due to an increase in the discharged stream temperature and pressure from the compressor, the energy consumption of this process was slightly increased compared with the proposed case. The sum of exergy input to the economical case (compressor-1) was 0.682 MW, which was 24.4 % of the conventional case, and the sum of energy input was 1.863 MW, which was 60.1 % of the conventional case.

As with other industrial distillation processes, Kansha et al. (2010c) developed azeotropic distillation for bioethanol production based on SHR. In order to separate pure ethanol from ethanol–water mixtures by distillation, it is necessary to use an entrainer (azeotroping agent) because the azeotropic mixture is one that vaporizes without any change in composition in azeotropic points. Thus, entrainers are used for this separation. Therefore, at least two separation units are required to produce pure ethanol, leading to increased energy consumption. Applying SHRT to this process, there was a reduction in the energy requirement to 1/8 of that of the conventional process. Also, Kansha et al. (2011) investigated the potential of further energy saving of the well-known and recently developed energy saving distillation HIDiC based on SHR. In HIDiC, the distillation column can be divided

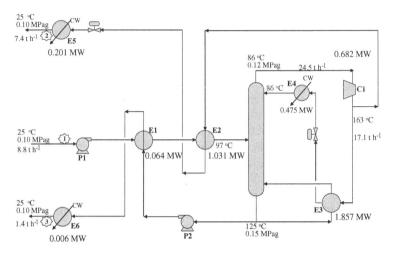

Fig. 3.3 Economical case

into two sections (the rectification section and the stripping section) and the condensation heat is exchanged with the vaporization heat between these two sections using the pressure difference.

Furthermore, the new design of crude oil distillation, the main factor of the energy consumption in a refinery plant, has been developed based on SHR (Kansha et al. 2012) and the energy saving efficiency compared with the conventional heat recovery with preheating heat exchanger network was investigated. In this new design, all the heat-related functions are pointed out and the conventional crude oil distillation unit is divided into the standardized modules by following SHR. These standardized modules are finally integrated.

3.8 Conclusion

In the commonly used type of distillation process (conventional case) based on the actual component balance for an industrial application, an external input (2.503 MW) is supplied to the reboiler and, simultaneously, the same amount of heat duty (2.503 MW) of the overhead vapor is disposed as waste heat. In contrast, by introducing SHRT in the proposed case, the overhead vapor is compressed and the enthalpy of the overhead vapor is increased due to adiabatic compression. This increased enthalpy can be fully utilized for both the reboiler duty and the feed heater duty. Eventually, the heat of the overhead vapor is recirculated in the process without the need for any external heat input. The exergy input in proposed case can be 21 % of that in the conventional case. Furthermore, the economical case was developed to increase the project cost-performance for industrial application by introducing single compressor system. The exergy input in the economical case can be 24 % of that in the conventional case. The economical case could not be reduced to the minimal exergy input, but it was considered that there was enough energy saving from the conventional case.

According to these case studies, and in spite of the difficulty of further reduction in energy consumption of a distillation process, SHRT was able to reduce the energy consumption considerably in the industrial application.

References

Annakou O, Mizsey P (1995) Rigorous investigation of heat pump assisted distillation. Heat Recovery Syst CHP 15:241–247

Aspelund A, Berstad DO, Gundersen T (2007) An extended pinch analysis and design procedure utilizing pressure based exergy for subambient cooling. Appl Therm Eng 27:2633–2649

Asprion N, Rumpf B, Gritsch A (2011) Work flow in process development for energy efficient processes. Appl Therm Eng 31(13):2067–2072

Brousse E, Claudel B, Jallut C (1985) Modeling and optimization of the steady state operation of vapor recompression distillation column. Chem Eng Sci 40(11):2073–2078

Campbell JC, Wigal KR, Van BV, Kline RS (2008) Comparison of energy usage for the vacuum separation of acetic acid/acetic anhydride using an internally heat integrated distillation column (HIDiC). Sci Technol Sep 43:2269–2297

Ceylan I, Aktas M, Dogan H (2007) Energy and exergy analysis of timber dryer assisted heat pump. Appl Therm Eng 27:216–222

Dhole VR, Linnhoff B (1992) Total site targets for fuel, co-generation, emissions, and cooling. Comput Chem Eng 7(Suppl 1):s101–s109

Enweremadu C, Waheed A, Ojediran J (2009) Parametric study of an ethanol–water distillation column with direct vapour recompression heat pump. Energy Sustain Dev 13(2):96–105

Ettouney H (2006) Design of single-effect mechanical vapor compression. Desalination 190:1–5

Fehlau M, Specht E (2000) Optimization of vapor compression for cost savings in drying processes. Chem Eng Technol 23:901–908

Fonyo Z, Benko N (1996) Enhancement of process integration by heat pumping. Comput Chem Eng 20:S85–S90

Gong G, Zeng W, Wang L, Wu C (2008) A new heat recovery technique for air-conditioning/heat-pump system. Appl Therm Eng 28:2360–2370

Haelssig JB, Tremblay AY, Thibault J (2008) Technical and economical considerations for various recovery schemes in ethanol production by fermentation. Ind Eng Chem Resour 47:6185–6191

Hirata K (2009) Heat integration of distillation column. Chem Eng Trans 18:39–44

Hou S, Zhang H (2009) An open reversed Brayton cycle with regeneration using moist air for deep freeze cooled by circulation water. Int J Therm Sci 48:218–223

Hou S, Li H, Zhang H (2007) Open air-vapor compression refrigeration system for air conditioning and hot water cooled by cool water. Energy Convers Manag 48:2255–2260

Jogwar SS, Daoutidis P (2009) Dynamics and control of vapor recompression distillation. J Process Control 19(10):1737–1750

Kansha Y, Tsuru N, Sato K, Fushimi C, Tsutsumi A (2009) Self-heat recuperation technology for energy saving in chemical processes. Ind Eng Chem Res 48(16):7682–7686

Kansha Y, Tsuru N, Fushimi C, Shimogawara K, Tsutsumi A (2010a) An innovative modularity of heat circulation for fractional distillation. Chem Eng Sci 65(1):330–334

Kansha Y, Tsuru N, Fushimi C, Tsutsumi A (2010b) Integrated process module for distillation processes based on self-heat recuperation technology. J Chem Eng Jpn 43(6):502–507

Kansha Y, Tsuru N, Fushimi C, Tsutsumi A (2010c) New design methodology based on self-heat recuperation for production by azeotropic distillation. Energy Fuels 24(11):6099–6102

Kansha Y, Kishimoto A, Tsutsumi A (2011) Process design methodology for high energy saving HIDiC based on self-heat recuperation. Asia Pac J Chem Eng 6(3):320–326

Kansha Y, Kishimoto A, Tsutsumi A (2012) Application of the self-heat recuperation technology to crude oil distillation. Appl Therm Eng (in press). doi:10.1016/j.applthermaleng.2011. 10.022

Klemes J, Dhole VR, Raissi K, Perry SJ, Puigjaner L (1997) Targeting and design methodology for reduction of fuel, power and CO_2 on total sites. Appl Therm Eng 7:993–1003

Matsuda K, Kawazuishi K, Kansha Y, Fushimi C, Nagao M, Kunikiyo H, Tsutsumi A (2011) Advanced energy saving in distillation process with self-heat recuperation technology. Energy 36:4640–4645

Nafey AS, Fath HES, Mabrouk AA (2008) Thermoeconomic design of a multi-effect evaporation mechanical vapor compression (MEE-MVC) desalination process. Desalination 230:1–15

Pavlas M, Stehlík P, Oral J, Klemes J, Kim JK, Firth B (2010) Heat integrated heat pumping for biomass gasification processing. Appl Therm Eng 30(1):30–35

Perry S, Klemes J, Bulatov I (2008) Integrating waste and renewable energy to reduce the carbon footprint of locally integrated energy sectors. Energy 33(10):1489–1497

Raissi K (1994) Total site integration. PhD thesis, UMIST, UK

Smith JM, Van Ness HC, Abbott MM (2005) Introduction to chemical engineering thermodynamics, 7th edn. McGaw-Hill, New York

Suphanit B (2010) Design of internally heat-integrated distillation column (HIDiC): uniform heat transfer area versus uniform heat distribution. Energy 35(3):1505–1514

Tarnawski VR, Leong WH, Momose T, Hamada Y (2009) Analysis of ground source heat pumps with horizontal ground heat exchangers for northern Japan. Renew Energy 34:127–134

Van der Ham LV, Kjelstrup S (2010) Exergy analysis of two cryogenic air separation processes. Energy 35:4731–4739

Wang L, Kano M, Hasebe S (2009) Effect of multiple steady-states on operation strategy and control structure for a heat integrated distillation column (HIDiC). Comput Aided Chem Eng 26:447–451

Wu C, Chen L, Sun F (1998) Optimization of steady flow heat pumps. Energy Convers Manag 39(5/6):445–453

Chapter 4
Drying Section

Abstract A large amount of energy is used in conventional methods for drying. But a novel drying process based on self-heat recuperation (SHR) is proposed to reduce energy consumption. A fluidized bed dryer with internal heat exchangers was selected as a main dryer and *Pinus radiata* wood chips were selected as a biomass sample. The mass and heat balance of a conventional drying process with heat recovery, and this proposed drying process, was calculated using a commercial process simulator PROII (Invensys, Inc.). The results show that the proposed drying process based on SHR can greatly reduce the energy consumption. In addition, this proposed drying process can appreciably decrease primary energy consumption and CO_2 emission during drying compared with the conventional drying process with heat recovery.

Keywords Drying · Fluidized bed · Vaporization · Water · Steam · Energy consumption · CO_2 emission · Sensible heat · Latent heat

4.1 Introduction

Carbonaceous material such as low-grade coal, woody/herbaceous/wet biomass, sludge, and manure usually contain large amounts of moisture (around 50–90 wt % in wet basis (wb)). A drying process is required (i) to reduce its cost for transportation, (ii) to increase thermal efficiency of thermochemical conversion processes, and (iii) to suppress decay during storage. Because the latent heat is large for evaporation of water (2.48–2.57 MJ (kg-evaporated $H_2O)^{-1}$ at an ambient temperature of 15 °C, depending on the wet bulb temperature [as reported by Brammer and Bridgwater (1999)], a large amount of energy is necessary for drying (Karthikeyan et al. 2009; Syahrul et al. 2002, 2003). Hence, an energy efficient drying technology needs to be developed to mitigate the environmental impact.

K. Matsuda et al., *Advanced Energy Saving and its Applications in Industry*,
SpringerBriefs in Applied Sciences and Technology,
DOI: 10.1007/978-1-4471-4207-2_4, © The Author(s) 2013

The intensification of heat and mass transfer in a dryer and the efficient utilization of energy including energy recovery in the dryer are both important for the improvement of the overall energy efficiency of drying process. The intensification of heat and mass transfer can decrease energy consumption because it optimizes heat supply, process time, and dryer configurations (Strumillo et al. 2006). Several technologies have been proposed to utilize the energy efficiently in a dryer. For example, heat recovery with flue gas recirculation (Iguaz et al. 2002; Ståhl and Berghel 2008) and without flue gas recirculation (Ogulata 2004), heat pump system (Krokida and Bishar 2004; Minea 2008), and pinch technology (Kemp 2005). Usually only the sensible heat is utilized in these drying technologies to raise the temperature for drying, and the latent heat of evaporated water/steam is not recovered. Because the latent heat of water/steam is much larger than the sensible heat of water, utilization of the heat of condensation is preferable to improve the overall thermal efficiency. From the view point of the utilization of heat of condensation, several researchers have proposed mechanical vapor recompression (VRC) technologies (Allardice et al. 2004; Aly 1999; Fehlau and Specht 2000; Hino 2003, 2005; Karthikeyan et al. 2009; Miller 1977; Nassikas and Akritidis 1992; Shibata and Mujumdar 1994). In VRC, a dryer compresses the exhausted steam and utilized its latent heat as a heat source for subsequent drying of a sample. However, these systems do not effectively recover all the heat of the evaporated water, the drying medium, or the dry products.

Recently, the self-heat recuperation technology (SHRT) has been applied to recuperate heat of water/steam in a drying process and it was determined that the energy consumption can be greatly reduced to approximately one-third to one-sixth of that required for conventional drying processes (Fushimi et al. 2011; Aziz et al. 2011). In this section, a concept of drying process by using SHRT is explained and energy consumption is calculated by using a commercial process simulator.

4.2 Basic Concept of Drying Based on Self-Heat Recuperation Technology

Figure 4.1 shows a basic concept of the proposed drying process based on self-heat recuperation (SHR) and its temperature diagram. The solid, dashed, and dotted lines show the flows of water including moisture in a solid sample, the sample, and the gas, respectively. The dryer consists of three parts; dryer 1: the sensible heat of water is exchanged for preheating, dryer 2: the latent heat of water and steam is exchanged for evaporation, dryer 3: the sensible heat of steam is exchanged for superheating.

Water in the wet sample in dryer 1 is preheated from room temperature (T_1) to its boiling point (T_2: around 100 °C) and evaporated in dryer 2. The evaporated water (steam) is then superheated to T_3 in dryer 3. The superheated steam is separated from the sample in a separator and is compressed by a compressor to raise its temperature for heat exchange (from $T_3{}'$ to $T_3{}'$). The sensible heat of the

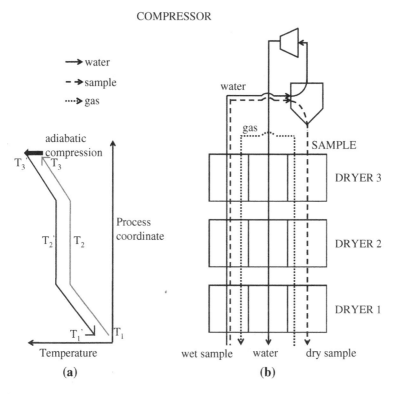

Fig. 4.1 **a** Basic concept of drying process based on self-heat recuperation, and **b** its temperature diagram

compressed superheated steam is exchanged for superheating of evaporated water in dryer 3. This results in a reduction of temperature of the compressed steam from T_3' to T_2'. The condensation heat of the compressed steam at T_2' is then exchanged to evaporate water from the sample in dryer 2. The sensible heat of the hot water is also recovered for preheating the wet sample in dryer 1. Eventually, the temperature of water decreased to T_1' and the water is exhausted. Furthermore, the sensible heat of the hot dry sample in this process is recovered by using gas, and the sensible heat is used to preheat sample to increase the overall thermal efficiency further. The dry sample is then obtained. In principle, this process can also be applied to drying at reduced pressure.

4.3 Drying Process

A novel drying process was designed based on SHR; Fig. 4.2 shows its schematic image. The flows of water including moisture in the wet and dried sample, sample with no moisture, and gas are represented as solid, dashed, and dotted lines, respectively. The wet sample (w_1 and b_1) is heated by warm water (w_7) and gas (g_7) in a heat exchanger (HX1) in dryer 1. Gas (g_1) is also heated in HX4. The preheated wet sample (w_2 and b_2) and gas (g_2) are fed into dryer 2. In dryer 2, the heat for evaporation is supplied by compressed steam (w_6) and gas (g_6) in HX2. The hot dry sample (b_3) is separated from gas and evaporated water and cooled by the gas (g_{11}) in HX3 resulting in a dry sample (b_4) being obtained. The evaporated steam (w_3) and gas (g_3) are superheated (w_4 and g_4) in HX5 and compressed by a compressor (w_5 and g_5). The sensible heat of the compressed steam and gas is exchanged in dryer 3 ($w_5 \rightarrow w_6$; $g_5 \rightarrow g_6$). The latent heat of the steam is then exchanged in dryer 2 ($w_6 \rightarrow w_7$; $g_6 \rightarrow g_7$). Subsequently, the sensible heat of warm water and gas is recovered in HX1 and HX4 in the dryer 1 ($w_7 \rightarrow w_8$; $g_7 \rightarrow g_8$). The mixture of water and gas are cooled in a condenser to separate condensed water (w_9), and the gas (g_{10}) with water vapor (w_{10}). The pressure energy of the gas (g_{10}) with water vapor (w_{10}) is partially recovered in an expander and the gas (g_{10}) with water vapor (w_{11}) is cooled in another condenser to remove condensed water (w_{12}). The gas (g_{11}) recovers the sensible heat of the hot dry sample (b_3). If a small amount of gas dissolves in water in the condensers, make up gas (g_{13}) is supplied to compensate for the loss. The gas ($g_{12} + g_{13} = g_1$) is then fed to HX4 and the condensed water (w_9 and w_{12}) is drained. Note that a larger amount of gas can promote evaporation, resulting in reduction of drying time and/or dryer size. However, excessive gas requires more compression energy and, thus, the amount of gas should be optimized.

Table 4.1 compares various types of dryers. In this study, a fluidized bed dryer (FBD) was selected because (i) fluidized beds have large contact surface area between solids and gas, (ii) the heat capacity coefficient is larger than other dryers, which results in smaller dryer size, and (iii) a gas, which recovers the sensible heat of the sample, can be used for fluidizing gas.

Pinus radiata wood chips are selected as the drying sample because of its various applications such as bioethanol and pulp productions (Araque et al. 2008; Aziz et al. 2011; Frederick et al. 2008; San Martín et al. 1995). However, it is usually difficult to fluidize wood chips owing to their peculiar shapes. Thus, a fine and inert medium is added to the fluidized bed dryer to promote the fluidity and drying rate of the wood chips (Aziz et al. 2011).

4.4 Mass and Heat Balance Calculation of the Proposed Drying Process

The mass and heat balance of the proposed drying process was calculated and the results were compared with that for a conventional drying process with heat recovery. Process simulation was conducted using a commercial process simulator PRO/II^{TM} Ver. 9 (Invensys) under the following conditions:

(1) the flow rate of the wet sample was 5,000 kg h^{-1}
(2) moisture content of the wet sample was 50 wt% on wet basis

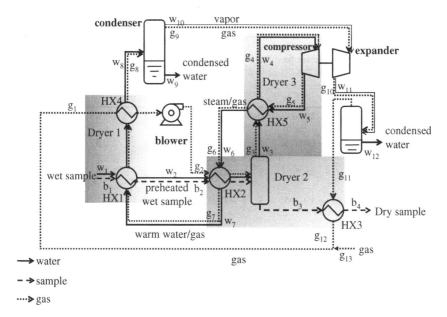

Fig. 4.2 Schematic image of the drying process based on self-heat recuperation

(3) moisture content of the dry sample was 20 wt% on wet basis (i.e., 75 % of water was removed)

(4) air was used as the gas

(5) dryer 2 with an internal heat exchanger consisted of a mixer, a heat exchanger, and a separator

(6) the minimum approach temperature of the heat exchanger was 30 K

(7) heat exchange was co-current in the dryer 2 and counter current in other heat exchangers

(8) the adiabatic efficiencies of a compressor, a blower, and an expander were 80 %, 80 %, and 90 %, respectively

(9) the pressure inside the FBD and the exhausted vapor from the FBD were atmospheric pressure (101.33 kPa)

(10) ambient temperature was 20 °C

(11) heat loss was neglected

Other conditions are listed in Table 4.2.

Figure 4.3a shows a schematic diagram and simulation results for the conventional heater drying process with heat recovery as a benchmark. The flows of water and sample are represented by solid and dashed lines, respectively. No air was used in this process. The regular value represents the energy input into the process, and italic value represents heat duty in a heat exchanger.

As can be seen, heat (1,267.0 kW) was input to the wet sample, which is a mixture of biomass (2,500 kg h^{-1}) and water (2,500 kg h^{-1}) in a heater (HEAT1). 1,875 kg h^{-1} of steam was then separated in a separator (SEP1) from the mixture

Table 4.1 Comparison of dryers (Society of Chemical Engineers, Japan 1999, 2011; Horio and Mori 1999)

Dryers	Heat capacity coefficient ha $(W\ m^{-3}\ K^{-1})$ or heat transfer coefficient $U\ (W\ K^{-1}\ m^{-2})$	Inlet hot air temperature or surface temperature $T\ (°C)$	Thermal efficiency (%)	Critial moisture content (%)	Samples
Fluidized bed dryer (convection)	$ha = 2,000–7,000$	100–600	50–65	1–10	Granules
Rotary dryer (convection)	$ha = 110–230$	200–600	40–70	2–10	Granules, Pellets,
Rotary dryer (conduction)	$U = 20–50$	100–150	60–75	2–10	Granules
Paddle dryer (convection)	$ha = 350–950$	400–800	60–70	2–3	Granules, Pellets, Sludge, Pastes
Paddle dryer (conduction)	$U = 80–350$	100–150	70–85	2–3	Sludges, Pastes, Cakes, Frakes

of biomass (2,500 kg h^{-1}) and remaining water (625 kg h^{-1}). Part of the sensible and latent heat of steam (114.9 kW) was recovered in the heat exchanger (HX1). The total energy consumption W_{base} was 1,267.0 kW.

Next, the mass and energy balance of the drying process based on SHR was calculated. The total energy consumption for this process, W_{pp}, can be calculated from the following equation:

$$W_{pp} = W_{cp} + W_{bl} - W_{ex}, \tag{4.1}$$

where W_{cp} is energy input to the compressor, W_{bl} is blower work, and W_{ex} is the recovered work by the expander, respectively. Note that W_{bl} was defined as the required work to increase the pressure for fluidization, P_f, and was calculated using the equations proposed by Kunii and Levenspiel (1991):

$$\Delta P_f = \Delta P_b + \Delta P_d \tag{4.2}$$

$$\Delta P_b = (1 - \varepsilon_{mf})(\rho_p - \rho_g)Hg \tag{4.3}$$

$$\Delta P_d = 0.4\Delta P_b \tag{4.4}$$

where ΔP_b and ΔP_d are the pressure drops across the bed and distributor, respectively, ε_{mf} is voidage at minimum fluidization condition, ρ_p is particle density, ρ_g is gas density, H is bed height, and g is gravitational acceleration.

Table 4.2 Material and bed properties used in process calculations (Aziz et al. 2011)

Properties	Values	Note and source
Inert particles (silica sand)		
Average diam. d_p (mm)	0.3	
Voidage at minimum fluidization ε_{mf}	0.42	Kunii and Levenspiel (1991)
Particle density ρ_p (kg m^{-3})	2,600	
Heat capacity C_p (kJ kg^{-1} K^{-1})	1.1	
Min. fluid. velocity U_{mf} (m s^{-1})	0.041	From experiment
Sample (*Pinus radiata*)		
Bulk density $\rho_{s,bulk}$ (kg m^{-3})	185	Dried condition
Particle density ρ_s (kg m^{-3})	600	
Heat capacity C_s (kJ kg^{-1} K^{-1})	$0.1031 + 0.003867T$	Simpson and TenWolde (1999)
Bed		
Shape	Square	
Side length y (m)	2	
Bed height H (m)	2	

Figure 4.3b shows the calculation results of the drying process based on SHR. The flows of water, sample, and gas are represented by solid, dashed, and dotted lines, respectively. The temperature, pressure, and mass flow rate of each stream are also shown. The regular values mean the energy input into the process and italic values mean heat duty in each heat exchanger (same as in Fig. 4.3a). The negative value of the expander (EX) represents the recovered energy. Note that the fraction of steam in gas is far below saturation point under this condition. Thus, the dryer 3 was omitted in this calculation because condensed water does not exist in the stream after the compression.

The wet sample (biomass 2,500 kg h^{-1} and water 2,500 kg h^{-1}) was preheated to 82.2 °C in a heat exchanger (HX1) and fed to the fluidized bed dryer. Heat duty in the HX1 was 180.9 kW. Air (3,044 kg h^{-1}) was also preheated in HX4, compressed in the blower (BL), and used as a fluidizing gas. The blower work, W_{bl}, was 38.1 kW. The heat of condensation in the compressed steam/air is used in HX2 to heat wet biomass and fluidizing air before drying of the sample. The heat duty of HX2 was 1,170.0 kW. The evaporated water and fluidizing gas (air) were separated from the dry biomass and compressed to 338 °C and 546 kPa by the compressor (CP). The compression work, W_{cp}, was 469.0 kW. The latent heat of the compressed steam/gas was recuperated in HX2 and its sensible heat was recovered in HX1 and HX4 in series, reducing its temperature to 101.7 °C. The mixture was fed into a flash drum (FL) to remove condensed water (1,463 kg h^{-1}) and part of the pressure energy was recovered by an expander ($W_{ex} = 142.4$ kW). The remaining water in the gas (412 kg h^{-1}) was removed in a condenser (CD). The cooled air recovered part of sensible heat of the hot dry sample (29.3 kW) in HX3, leading to the decrease in temperature of the dry sample from 90 to 50 °C. The total energy consumption of this process, W_{pp}, was:

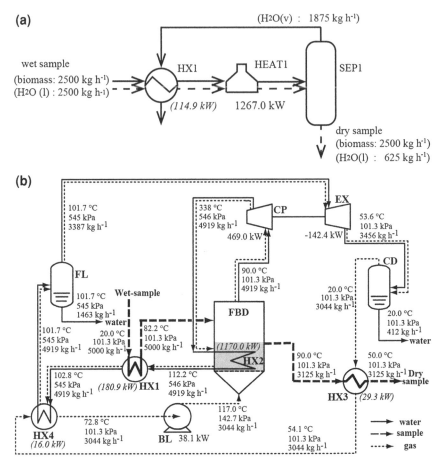

Fig. 4.3 a Simulation results of conventional drying process with heat recovery; **b** simulation results of the proposed drying process

$$W_{pp} = W_{cp} + W_{bl} - W_{ex} = 469.0 + 38.1 - 142.4 = 364.7\,\text{kW} \qquad (4.5)$$

This value is 28.8 % (= 364.7/1,267.0) of the energy consumption for the conventional drying process with heat recovery (W_{base}) and corresponds to 700.2 kJ (kg-H_2O evaporated)$^{-1}$, which was much smaller than the latent heat of water.

$$(= 364.7\,\text{kW} \times 3,600\,\text{s}\,\text{h}^{-1}/1,875(\text{kg - }H_2O\,\text{evaporated})\,\text{h}^{-1}) \qquad (4.6)$$

zThus, it can be said that the recuperation of both sensible and latent heats can significantly increase the efficiency of the dryer, resulting in energy savings for drying.

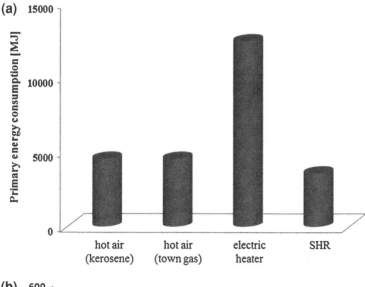

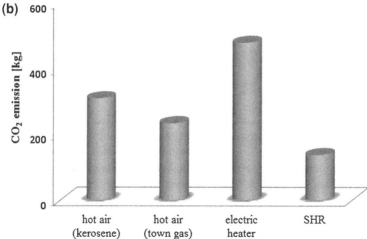

Fig. 4.4 **a** Required primary energy to dry 5,000 kg h^{-1} of biomass from 50 wt% (wet basis) to 20 wt% (wet basis); **b** CO_2 emission to dry 5,000 kg h^{-1} of biomass from 50 wt% (wet basis) to 20 wt% (wet basis)

4.5 Primary Energy Consumption and CO_2 Emission

Based on the results in the previous section, primary energy consumption and CO_2 emission to dry 5,000 kg h^{-1} of biomass from 50 wt% (wet basis) to 20 wt% (wet basis) was calculated by using the following assumptions:

(1) Kerosene: thermal efficiency of a boiler was 1.0 and CO_2 emission factor was 0.0679 kg-CO_2 MJ^{-1} (Ministry of the Environment Government of Japan 2005)
(2) Town gas: thermal efficiency of a boiler was 1.0 and CO_2 emission factor was 0.0513 kg-CO_2 MJ^{-1} (Ministry of the Environment Government of Japan 2005)
(3) Electricity: net thermal efficiency of electric power generation was 36.6 % and CO_2 emission factor was 0.378 kg-CO_2 kWh^{-1} (Ministry of the Environment Government of Japan 2005)

Figure 4.4a, b shows the calculated primary energy consumption and CO_2 emission, respectively. The drying process based on SHR can reduce primary energy consumption to 78.6 % (=3587.2 MJ/4561.2 MJ) compared with the conventional drying process, even if the required electricity is converted into primary energy by using the net thermal efficiency of electric power generation. By comparing the drying with an electric heater, the primary energy consumption of the SHR drying process can be reduced to 28.8 % (=3578.2 MJ/12462.3 MJ). In addition, the CO_2 emission of the drying process based on SHR is reduced to 44.5 % (=137.9 kg-CO_2/309.6 kg-CO_2) of that from a kerosene boiler, or 58.9 % (=137.9 kg-CO_2/234.0 kg-CO_2) of town gas boiler drying process. These results indicate the potential of the drying process based on SHR to greatly reduce not only primary energy consumption but also CO_2 emission.

4.6 Conclusion

A novel drying process based on SHR was proposed to effect a significant reduction in energy consumption. The mass and energy balance of the process with a fluidized bed type dryer was calculated using commercial simulation software (PROII version9, Invensys), which showed that the total energy input of the SHR process was 364.7 kW, which was only 28.8 % of the required energy for a conventional heater drying process with heat recovery (1267.0 kW). In addition, it was determined that the drying process based on SHR greatly reduced primary energy consumption and CO_2 emission compared with the conventional drying process.

References

Allardice DJ, Chaffee AL, Jackson WR, Marshall M (2004) Water in brown coal and its removal. In: Li CZ (ed) Advances in the science of Victorian brown coal. Elsevier, Amsterdam
Aly SE (1999) Energy efficient combined superheated steam dryer/MED. Appl Therm Eng 19:659–668
Araque E, Parra C, Freer J, Contreras D, Rodríguez J, Mendonça R, Baeza J (2008) Evaluation of organosolv pretreatment for the conversion of *Pinus radiata* D. Don to ethanol. Enzyme Microb Technol 43:214–219

Aziz M, Fushimi C, Kansha Y, Mochidzuki K, Kaneko S, Tsutsumi A, Matsumoto K, Hashimoto T, Kawamoto N, Oura K, Yokohama K, Yamaguchi Y, Kinoshita M (2011) Innovative energy-efficient biomass drying based on self-heat recuperation. Chem Eng Technol 34:1095–1103

Fehlau M, Specht E (2000) Optimization of vapor compression for cost savings in drying processes. Chem Eng Technol 23:901–908

Frederick WJ, Lien SJ, Courchene CE, DeMartini NE, Ragauskas AJ, Iisa K (2008) Co-production of ethanol and cellulose fiber from Southern Pine: a technical and economic assessment. Biomass Bioenergy 32:1293–1302

Fushimi C, Kansha Y, Aziz M, Mochidzuki K, Kaneko S, Tsutsumi A, Matsumoto K, Kawamoto N, Oura K, Yokohama K, Yamaguchi Y, Kinoshita M (2011) Novel drying process based on self-heat recuperation technology. Dry Technol 29:105–110

Hino T (2003) Energy-saving drying technology by vapor recompression. Kagaku-souchi 45:45–49 (in Japanese)

Hino T (2005) Possibility of VRC dehydration for energy production from wet biomass. J Jpn Inst Energy 84:353–358 (in Japanese)

Horio M, Mori S (eds) (1999) The Association of Powder Process Industry and Engineers, Japan. In: Handbook of fluidization. Baifukan, Tokyo

Iguaz A, Lopez A, Virseda P (2002) Influence of air recycling on the performance of a continuous rotary dryer for vegetable wholesale by-products. J Food Eng 54:289–297

Karthikeyan M, Wu Z, Mujumdar AS (2009) Low-rank coal drying technologies-current status and new developments. Dry Technol 27:403–415

Kemp IC (2005) Reducing dryer energy use by process integration and pinch analysis. Dry Technol 23:2089–2104

Krokida MK, Bishar GI (2004) Heat recovery from dryer exhaust air. Dry Technol 22:1661–1674

Kunii D, Levenspiel O (1991) Flulidization engineering, 2nd edn. Butterworth-Heinemann, Oxford

Miller W (1977) Energy conservation in timber-drying kilns by vapor recompression. For Prod J 27:54–58

Minea V (2008) Energetic and ecological aspects of softwood drying with high-temperature heat pumps. Dry Technol 26:1373–1381

Ministry of the Environment Government of Japan (2005) Guidelines of CO_2 emission factors (in Japanese). http://www.env.go.jp/earth/ondanka/santeiho/guide/index.html. Accessed 27 Dec 2011

Nassikas AA, Akritidis CB (1992) Drying heat pump with water vapour as working medium. Dry Technol 10:239–250

Ogulata RT (2004) Utilization of waste-heat recovery in textile drying. Appl Energy 79:41–49

San Martín R, Perez C, Briones R (1995) Simultaneous production of ethanol and kraft pulp from pine (*Pinus radiata*) using steam explosion. Bioresour Technol 53:217–223

Shibata H, Mujumdar AS (1994) Steam drying technologies: Japanese R&D. Dry Technol 12:1485–1524

Simpson W, TenWolde A (1999) Wood handbook-wood as an engineering material. Forest Product Laboratory, USDA Forest Service, Wisconsin

Society of Chemical Engineers, Japan (1999) Handbook of chemical engineering, 6th edn. Maruzen, Tokyo, Chapter 14 (in Japanese)

Society of Chemical Engineers, Japan (2011) Handbook of chemical engineering, 7th edn. Maruzen, Tokyo, Chapter 5 (in Japanese)

Ståhl M, Berghel J (2008) Validation of a mathematical model by studying the effects of recirculation of drying gases. Dry Technol 26:786–792

Strumillo C, Jones PL, Zylla R (2006) Energy aspects in drying. In: Mujumdar AS (ed) Handbook and industrial drying. Taylor & Francis, Boca Raton

Syahrul S, Hamdullahpur F, Dincer I (2002) Exergy analysis of fluidized bed drying of moist particles. Appl Therm Eng 22:1763–1775

Syahrul S, Dincer I, Hamdullahpur F (2003) Thermodynamic modeling of fluidized bed drying of moist particles. Int J Therm Sci 42:691–701

Brammer JG, Bridgwater AV (1999) Drying technologies for an integrated gasification bio-energy plant. Renew Sustain Energy Rev 3:243–289

Chapter 5
Gas Separation Section

Abstract In this chapter, self-heat recuperation technology (SHRT) is applied in gas separation processes, which are (1) cryogenic air separation and (2) CO_2 chemical absorption processes. (1) The energy consumption of cryogenic air separation based on self-heat recuperation (SHR) can be reduced by 40 % compared with a conventional process. In the proposed process, not only the latent heat but also the sensible heat of the process stream is circulated in the process. Furthermore, the pressure in the column can be reduced compared with the high pressure part of a conventional cryogenic air separation process. (2) The energy consumption of a CO_2 chemical absorption process based on SHR can be reduced by 70 % compared with the conventional process. In the proposed process, the heat of the exothermic absorption reaction in the absorber and the heat of the steam condensation in the condenser of the stripper are recuperated and circulated for reuse for regeneration of absorbent solution and vaporization in water during CO_2 stripping.

Keywords CO_2 chemical absorption · Cryogenic air separation · Energy saving · Gas separation · Post-combustion · Self-heat recuperation technology

5.1 Introduction

Gas separation processes are applied to a great variety of industries. Moreover, industrial gasses such as O_2, N_2, H_2, and CO are produced in a variety of separation processes, thus, it is important that the energy consumption of the gas separation process is reduced.

Cryogenic air separation and the CO_2 chemical absorption process based on SHR are mentioned in this section as typical examples. Cryogenic distillation is

K. Matsuda et al., *Advanced Energy Saving and its Applications in Industry*,
SpringerBriefs in Applied Sciences and Technology,
DOI: 10.1007/978-1-4471-4207-2_5, © The Author(s) 2013

generally applied to air separation and chemical absorption is generally applied to the CO_2 separation process. These gas separation processes utilize pressure and temperature swings for separation. Thus, if the energy of heat and pressure is recirculated and utilized in these processes, then the energy consumption can be reduced.

5.2 Cryogenic Air Separation (Distillation)

The cryogenic distillation can separate high purity O_2 and N_2 from air, but it is well known that cryogenic distillation requires a large amount of energy to liquefy air by cooling (Boehme et al. 2003). To produce liquid nitrogen in the cryogenic distillation process, a two-column separation process is applied. The two-column process consists of low and high pressure columns, which combine the condenser of the high pressure column with the reboiler of the low pressure column. As a result, the latent heat of vaporization of O_2 can be exchanged with the condensation heat of pure nitrogen in the two-column separation process (Kerry et al. 2006; Thiemens et al. 1984; Nakaiwa et al. 1996; Seliger et al. 2006; Miller et al. 2008; Zhu et al. 2009). Thus, not only is nitrogen liquefied in the process but the energy consumption of the process can also be decreased by this heat exchange. However, a large pressure difference between the high and low pressure columns is required to exchange the vaporization heat. Therefore, the cost of the energy needed to produce pure oxygen still remains quite high due to the work needed to compress the feed air (Roffel et al. 2000; Bian et al. 2005).

5.2.1 Conventional Cryogenic Air Separation Process

Figure 5.1 shows a flow diagram of the conventional cryogenic air separation process. This process can be divided into two sections, which are the heat exchange section and distillation section. In the heat exchange section, air (1) (25 °C, 1 atm) is fed into a compressor (Comp1), the stream 1 is then compressed by a compressor with cooling by an after-cooler (C1) $(1 \rightarrow Comp1 \rightarrow 2 \rightarrow C1 \rightarrow 3)$. The cooled stream 3 is fed into a main heat exchanger (HX1) to be cooled by the lower temperature effluent streams from the distillation section $(9 \rightarrow 10, 12 \rightarrow 13$ and $15 \rightarrow 16)$. The pressurized and cooled gaseous air (4) is fed into the cold box $(4 \rightarrow CB)$, which is the distillation section. The pressurized cooled air in the distillation section is separated into high purity product O_2 (9), low purity exhaust N_2 (12), and high purity product N_2 (15) by two distillation columns: lower column (LC) and upper column (UC). The condensation heat of the lower column is exchanged with vaporization heat of the upper column. Thus, the latent heat is recovered in the conventional process but this requires a large amount of energy for compression work in Comp1.

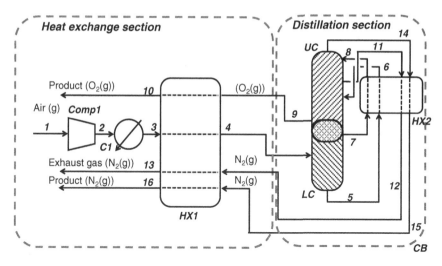

Fig. 5.1 Process configuration of the conventional cryogenic air separation process

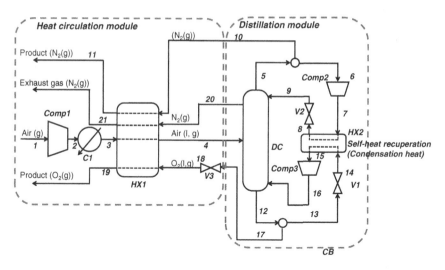

Fig. 5.2 Process configuration of the proposed cryogenic air separation process

5.2.2 Proposed Cryogenic Air Separation Process

Figure 5.2 shows a flow diagram of the proposed cryogenic air separation process. This process can also be divided into two sections: the heat circulation and the distillation modules based on SHR (see Chap. 1). By using SHRT, the total enthalpy of the inlet streams in each section is equal to the total enthalpy of the outlet streams.

In the heat circulation module based on SHR, air (1) is fed into a compressor and the stream 1 is compressed by a compressor (Comp1) and is cooled by an after-cooler (C1) (1→Comp1→2→C1→3). The cooled stream 3 is fed into a heat exchanger (HX1) to be cooled by the lower temperature effluent streams from the distillation module based on SHR.

The pressurized cooled air (4), which has gas and liquid mixed phases, is fed into the distillation module based on SHR. Stream 4 in this module is fed into the middle of the distillation column (DC). The top product of the DC (5) is divided into two streams; the high purity N_2 gas (10) and the reflux stream (6). Stream 6 is compressed by a compressor (Comp2) to recuperate the condensation heat in the heat exchanger (HX2) (6→Comp2→7→HX2→8). Subsequently, stream 8 is de-pressurized by a valve (V2) (8→V2→9). Stream 9 is 100 % liquid phase, and the stream 9 is fed into the DC as a reflux stream. Simultaneously, the bottom product (12) of the DC is divided into two streams; the high purity O_2 product stream 17 and stream 13. Stream 13 is depressurized by a valve (V1), and stream 14 is heated by condensation heat in the heat exchanger (HX2) (13→V1→14→HX2→15). Subsequently, the high purity product O_2 becomes a gaseous stream (15), which is compressed by compressor (Comp3) (15→Comp3→16). The compressed gaseous stream (16) is returned to the DC. Finally, the high purity N_2 gas (10), low purity N_2 gas (20), and high purity O_2 liquid (17) are separated from the feed air by the DC. In the heat circulation module based on SHR, the cold energy of these product streams (10, 17, 20) are utilized as cooling air (1) in HX1.

5.2.3 Mass and Heat Balance Calculation

A process simulation was conducted to elucidate the energy saving effect of the proposed air separation process compared with the conventional air separation process. The energy requirements were calculated for an ideal fluid stream by using the commercial simulator PRO/II^{TM} version 8.1 (Invensys). This model case of a cryogenic air separation process separates into high purity O_2 (99.99 mol %, 31,000 m^3 h^{-1} at standard temperature and pressure) and high purity nitrogen (99.9 mol %, 30,000 m^3 h^{-1} at standard temperature and pressure). The flow rate of the feed stream and its composition are assumed as 1,67,000 m^3 h^{-1}, 80 mol % N_2, and 20 mol % O_2, respectively. The number of stages of the UC, LC and DC are assumed as 20, 7 and 100, respectively. The real air consists of nitrogen (78.08 %), oxygen (20.95 %), argon (0.93 %), carbon dioxide (0.03 %) and etc. The energy efficiency of the separation process of N_2/O_2 mixture is simply compared on the same basis.

The ideal gas equation of state was applied. The adiabatic efficiency in the compressor was assumed 100 %. From the simulation results, the total work of the conventional cryogenic air separation process was 17,800 kW (574.2 kWh kNm^{-3}-O_2, Fig. 5.3). The total work is supplied to the compressor (Comp1) (1→Comp1→2), and the compressed stream (446 kPag) is cooled to 10 °C by an

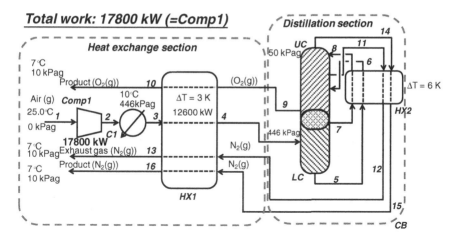

Fig. 5.3 Simulation results of the conventional cryogenic air separation process

after-cooler (C1) (2→C1→3). Moreover, the stream 3 is cooled with product streams in HX1 ($\Delta T = 3$ K, 12,600 kW). The feed pressure of the LC was 446 kPag and that of the UC was 50 kPag, to produce the liquid phase of the stream in the conventional column by latent heat exchange. These pressures of LC and UC depended on concentrations of the products and conditions of latent heat exchange between the top of LC and the bottom of UC.

In contrast, the total work of the proposed cryogenic air separation process based on SHR can be reduced by 11,350 kW (11,350 kW = 11000 + 150 + 200 kW, 366.1 kWh kNm^{-3}-O$_2$, Fig. 5.4). Feed stream 1 is compressed by the compressor (Comp1) (1→Comp1→2), and the compressed stream (200 kPag) is cooled to 10 °C by an after-cooler (C1) (2→C1→3). Although, the self-heat exchange duty increased to 15,000 kW in HX1 to cool the feed stream to −181 °C to balance the inlet and outlet stream enthalpy in the heat circulation module based on SHR, the compression duty, because of the lower stream pressure, decreased to 11,000 kW (Comp1). In the distillation module based on SHR, the feed pressure of the DC was 200 kPag, which was decided by the temperature condition distillation module based on SHR. In the same way, the pressure in conventional process is decided by the temperature condition. The pressure difference between the column (DC) of the feed stream based on SHR and the conventional columns (LC, UC) of feed stream indicates that whole process heat can be recirculated in the proposed distillation module based on SHR, and the pressure of the DC can be lower than that of LC. Therefore, the reflux ratio of the DC can be smaller than that of the LC, and the heat exchange duty to separate the product from air can be reduced. To recirculate the process heat in the distillation module based on SHR, compression work is required by compressors (Comp2: 150 kW, Comp3: 200 kW).

The simulation results indicated that the proposed cryogenic air separation process based on SHR could reduce energy consumption in the process by about

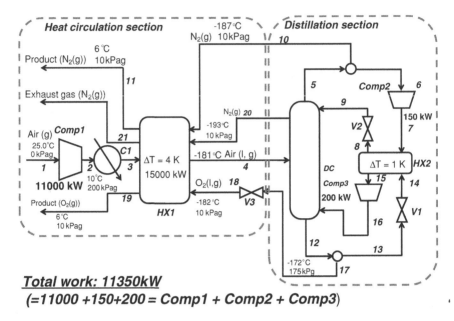

Fig. 5.4 Simulation results of the proposed cryogenic air separation process

40 % compared with the conventional cryogenic air separation process. It can be seen that the loss of work for compression is transformed into heat as additional heating for the process stream and recirculated into the process stream, which did not provide any significant effect to the required energy ratio between the proposed process and the conventional process.

5.3 CO_2 Absorption Process

Carbon capture and storage (CCS) has attracted significant attention in the past two decades as a means to reduce greenhouse gas emissions and mitigate global warming. CCS consists of the separation of CO_2 from industrial and energy-related sources, transportation of CO_2 to a storage location and long-term confinement of CO_2 away from the atmosphere (IPCC 2005; Rubin et al. 2005).

It has been observed that the most significant stationary point sources of CO_2 emissions are power generation processes. In fact, the CO_2 emissions from power generation processes comprise 40 % of the global CO_2 emissions (Rubin et al. 2005; Toftegaard 2010). There are three different types of CO_2 capture processes for power generation: post-combustion, pre-combustion, and oxy-fuel combustion (Rubin et al. 2005). The CO_2 absorption process for post-combustion is focused in this section.

Post-combustion capture in power plants is generally used for pulverized-coal-fired power plants. The CO_2 concentration in post-combustion is low compared

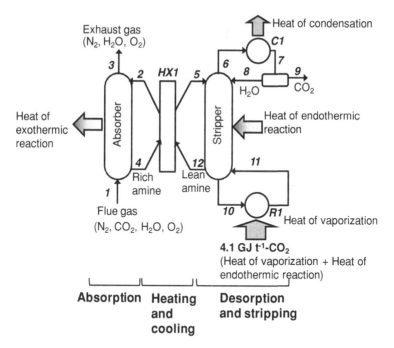

Fig. 5.5 Conventional CO$_2$ absorption process

with the other two CO$_2$ capture processes: around 10 % (wet base). The CO$_2$ capture is typically performed via chemical absorption with monoethanolamine (MEA), methyldiethanolamine (MDEA) and diethanolamine (DEA) absorbent. MEA is widely used in conventional CO$_2$ capture plants.

5.3.1 Conventional CO$_2$ Absorption Process

Figure 5.5 shows a diagram of the conventional CO$_2$ absorption process, which consists of an absorber, a heat exchanger (HX1) for heat recovery, and a stripper (regenerator) with a reboiler. The flue gas (1) and a 'lean CO$_2$ concentration' amine solution (lean amine) (2) are fed into the absorber, and CO$_2$ gas is absorbed into the lean amine.

This amine solution containing absorbed CO$_2$ is called the 'rich CO$_2$ concentration' amine solution (rich amine) (4). Exhaust gases are discharged from the top of the absorber (3). The rich amine is fed into the stripper through the HX (4→HX1→5) and lean amine is then regenerated and the CO$_2$ gas is stripped by heating in the reboiler (R1) of the stripper (10→R1→11). CO$_2$ gas and vapor stream from top of stripper is cooled by condenser (C1) (6→C1→7). Stream 7 is separated into CO$_2$ (9) and H$_2$O (8) streams and stream 8 is returned to the stripper.

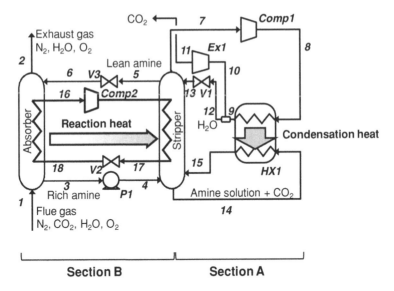

Fig. 5.6 CO$_2$ absorption process based on self-heat recuperation

The heat in the conventional absorption process using MEA (4.1 GJ t^{-1}-CO$_2$) is supplied by the reboiler in the stripper. The ratio of this heat for regeneration and vaporization is 1:1. From Fig. 5.5, it can be understood that a part of the sensible heat is recovered from lean amine using the HX1, because the temperature of lean amine is higher than that of rich amine. However, the heat of steam condensation cannot be recovered for heat of vaporization for stripping in the reboiler because of the temperature difference between the condenser and the reboiler. Thus, CO$_2$ capture is the most costly and the highest energy consumption process of power generation, leading to higher CO$_2$ emissions. In fact, some studies report that this process drops the net efficiency of the power plant by about 10 % (Damen 2006, Davison 2007).

If all process heat (sensible heat, latent and reaction heat) can be recirculated into the process, the energy required for CO$_2$ capture can be greatly reduced.

5.3.2 Proposed CO$_2$ Absorption Process

To achieve perfect internal heat circulation, SHRT was applied to the CO$_2$ absorption process as shown in Fig. 5.6 (Kishimoto et al. 2011).

In Fig. 5.6, the distillation module based on SHR (see Chap. 1) is able to apply to section A. Steam (7) is discharged from the top of the stripper and compressed adiabatically by a compressor (Comp1) to recuperate the steam condensation heat in HX1. This recuperated heat is exchanged with the heat of vaporization for stripping, leading to a reduction in the energy consumption for stripping.

The heat circulation module based on SHR (see Chap. 1) can be used in section B. Furthermore, the heat of the exothermic reaction generated at low temperature in the absorber is transported and utilized as reaction heat for solution regeneration at high temperature using a reaction heat transformer (RHT) (16→Comp2→17→V2→18). This RHT is a type of closed-cycle compression system with a volatile fluid as the working fluid and consists of an evaporator to receive heat from the heat of exothermic reaction in the absorber, a compressor, a condenser to supply heat to the stripper as heat of the endothermic reaction, and an expansion valve. The heat of the exothermic absorption reaction at the evaporator in the absorber is transported to the endothermic desorption reaction in the condenser of the stripper by the RHT. Therefore, both the heat of the exothermic absorption reaction in the absorber and the heat of steam condensation from the condenser in the stripper are recuperated and utilized as the reaction heat for solution regeneration and the vaporization heat for CO$_2$ stripping in the reboiler of the stripper.

5.3.3 Mass and Heat Balance Calculation

A process simulation was conducted using the standard amine package of PRO/II (INVENSYS PRO/II 8.1^{TM}) to compare the energy consumption between the proposed CO$_2$ gas separation process and the conventional process. It was assumed that the flow rates of flue gas is 100 kmol h^{-1}, and the flue gas consists of N$_2$ 71.1 %, CO$_2$ 12.4 %, H$_2$O 12.2 %, and O$_2$ 4.3 %. In this simulation, the temperature difference in heat exchanger was set to 10 K. Figure 5.7 shows the simulation result of the conventional absorption process.

The total input energy in this process is 634.3 kW (=0.3 + 634.0 = P1+R1), and some amount of the heat of the exothermic absorption reaction is recovered by rich amine. However, the exhoust gas (N$_2$, H$_2$O, O$_2$) takes some amount of heat of the exothermic absorption reaction from the absorber. Furthermore, a large amount of process heat is wasted in cooling water. The sum of cooler duty and condenser duty is 643.8 kW(=55.0 + 243.9 + 344.9 = C3 + C2 + C1). Although a part of the heat is recovered from the heat of lean amine in HX1 (441.8 kW), most of the reboiler duty (634.0 kW) and heat of exothermic reaction is wasted in cooling water.

Figure 5.8 shows simulation results of the proposed absorption process. In the stripping section (Section A), heat of condensation from the stream at the top of the stripper is utilized as heat of vaporization in the reboiler by compression work. The energy required in the stripping section is 103.3 kW (=110.3−7.0 = Comp1-Ex1), which includes recovered energy (−7.0 kW, Ex1). In order to regenerate amine solution in section B, the heat of the exothermic reaction is utilized as heat of endothermic reaction by compression work. The total energy consumption of the RHT was 98.8 kW (Comp2) when the coefficient of performance (COP) was set at 3. This COP(η) is expressed as Q/W. Q is the heat received from the heat of the exothermic absorption reaction [kW] and W is the compression work in the

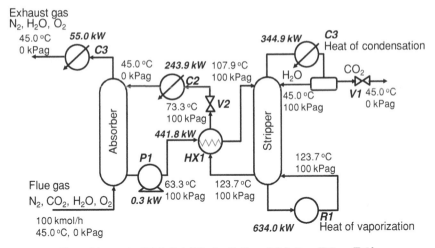

Total input: 634.3 kW (= 0.3 + 634.0 = P1 + R1)

Fig. 5.7 Simulation results of the conventional absorption process

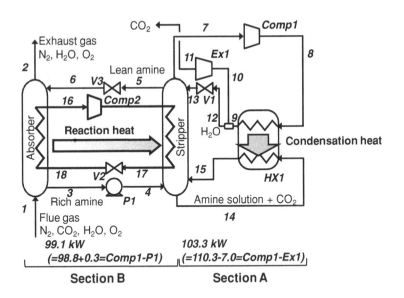

Fig. 5.8 Simulation results of the proposed absorption process

RHT [kW]. Thus, the total energy consumption is 202.4 kW (=110.3− 7.0 + 98.8 + 0.3 = Comp1-Ex1+Comp2+P1).

5.4 Conclusion

Two gas separation processes (cryogenic air separation process and CO_2 absorption process based on SHR) are proposed in this section. A cryogenic air separation process based on SHR is also proposed. By applying the SHRT to the cryogenic air separation process, not only the latent heat but also the sensible heat of the process stream can be circulated in the process. Simulation results indicated that the proposed cryogenic air separation process based on SHR could reduce energy consumption in the process by about 40 % compared with the conventional cryogenic air separation process. A novel process for CO_2 gas separation using chemical absorption based on SHR is proposed. The proposed CO_2 gas separation in post-combustion capture, in which not only the process heat but also the reaction heat is recovered and recirculated, results in a significant reduction in energy consumption compared with conventional processes. From the simulation results, the energy consumption of the proposed process can lead to decrease by 32 % (=202.4/634.3) of that consumed in the conventional self-heat recovery process when the COP of the RHT is set at 3.

References

Bian S, Khowinij S, Henson MA, Belanger P, Megan L (2005) Compartmental modeling of high purity air separation columns. Comp Chem Eng 29:2096–2109

Boehme R, Parise JAR, Marques RP (2003) Simulation of multistream plate-fin heat exchangers of an air separation unit. Cryogenis 43:325–334

Damen K, Troost VM, Faaij A, Turkenburg W (2006) A comparison of electricity and hydrogen production systems with CO_2 capture and storage. Part A: Review and selection of promising conversion and capture technologies progress in energy and combustion. Science 32(6):215–246

Davison J (2007) Performance and costs of power plants with capture and storage of CO_2. Energy 32:1163–1176

IPCC special report on carbon dioxide capture and storage (2005) Prepared by working group III of the intergovernmental panel on climate change (IPCC). Cambridge University Press, New York

Kerry FG (2006) Industrial gas handbook: gas separation and purification. CRC Press, Taylor and Francis, New York

Kishimoto A, Kansha Y, Fushimi C, Tsutsumi A (2011) Exergy recuperative CO_2 gas separation in post-combustion capture. Ind Eng Chem Res 50:10128–10135

Miller J, Luyben WL, Belanger P, Blouin S, Megan L (2008) Improving agility of cryogenic air separation plants. Ind Eng Chem Res 47:394–404

Nakaiwa M, Akiya T, Owa M, Tanaka Y (1996) Evaluation of an energy system with an air separation. Enery Convers Manag 37:295–301

Seliger B, Hanke-Rauschenbach R, Hannemann F, Sundmacher K (2006) Modelling and dynamics of an air separation rectification column as part of an IGCC power plant. Sep Purif Technol 49:136–148

Thiemens MH, Meagher D (1984) Cryogenic separation of nitrogen and oxygen in air for determination of isotopic ratios by mass spectrometry. Anal Chem 56:201–203

Toftegaard MB, Brix J, Jensen PA, Glarborg P, Jensen AD (2010) Oxy-fuel combustion of solid fuels. Prog Energy Combust Sci 36:581–625

Roffel B, Betlem BHL, de Ruijter JAF (2000) First principles dynamic modeling and multivariable control of a cryogenic distillation process. Comp and Chem Eng. 24:111–123

Rubin ES, Chen C, Rao AB (2007) Cost and performance of fossil fuel power plants with CO_2 capture and storage. Energy policy 35:4444–4454

Zhu L, Chen Z, Chen X, Shao Z, Qian J (2009) Simulation and optimization of cryogenic air separation units using a homotopy-based backtracking method. Sep Purif Technol 67:262–270

Part III
Utility System

Chapter 6
Utility System

Abstract The total site approach based on pinch technology, "Area-wide pinch technology", which consists of total site profile (TSP) analysis and R-curve analysis, was applied to two of the biggest heavy chemical complexes in Japan, i.e., Mizushima and Chiba industrial areas. This study demonstrated that there was a huge amount of energy saving potential through energy sharing among various sites despite the very high efficiency of the individual sites in the complexes. As a result, energy saving projects have been developed on the basis of the study.

Keywords Utility system · Total site approach · Area-wide · Pinch technology · Total site profile · R-curve

6.1 Introduction

Previous energy saving studies for both process and utility systems have been carried out based on the concept of a single site approach. The single site approach would improve the energy efficiency within the site itself. A utility system provides heat and power for all the process systems in a site and, for this reason, the concept of total site approach for energy saving study in utility system by pinch technology was developed. However, the application of this concept was still limited to a single site.

Recently it has often been surmised that all possible energy saving measures in heavy chemical sites within complexes had already been completed for study and implementation. In order to overcome this limitation and to achieve the further improvement, the expansion type of the total site approach "Area-wide pinch technology" was considered (Matsuda 2008). The utility systems which many sites possess would be better to be integrated totally. It was considered that energy

K. Matsuda et al., *Advanced Energy Saving and its Applications in Industry*,
SpringerBriefs in Applied Sciences and Technology,
DOI: 10.1007/978-1-4471-4207-2_6, © The Author(s) 2013

efficiency could be improved significantly by area-wide optimization in design and operation, under the concept of one large virtual site sharing heat and power.

Looking at the utilization of heat by the utilities in a site, there are a steam heaters and reboilers, using hot utilities for heating, and coolers and steam generators which use cold utilities for cooling. It is therefore possible to find a large temperature difference between heat supply side and heat demand side. It is often said that low-grade heat (around 150–200 °C) of process streams is cooled by coolers and disposed of as waste heat, however low-pressure steam (around 130 °C) could be produced from such heat. Middle-pressure steam (200–250 °C) is sometimes used for a reboilers but it can be replaced with the lower pressure steam, if the process stream requires the low level heat (about 110 °C). From above information, it would be possible to look at the energy saving potential of energy sharing (heat sharing). For this purpose, area-wide pinch technology is considered to be an appropriate methodology by which an energy saving study of a whole industrial area could be undertaken.

6.2 Area-Wide Pinch Technology

"Area-wide pinch technology" consists of total site profile (TSP) analysis and R-curve analysis.

6.2.1 Total Site Profile Analysis

In the context of the total site approach, whereby many process plants are integrated for optimization and energy saving, the utility system must be considered and optimized by a total site approach. A graphical method, so-called site profiles, was first introduced by Dhole and Linnhoff (1992) and later extended by Raissi (1994). Klemes et al. (1997) considerably extended this methodology to site-wide applications. Data for individual process heat recovery are firstly converted to grand composite curves (GCCs). GCCs are combined to form a site heat source profile and a site sink profile. These two profiles form total site profiles (TSP) analogous to the composite curves for individual processes. Perry et al. (2008) extended the site utility grand composite curves (SGCC). Bandyopadhyay et al. (2010) developed a methodology to estimate the cogeneration potential of an overall site through SGCC.

TSP analysis combines the heat supply and demand using the heat exchanger data. In Fig. 6.1, the right side of TSP shows the composite curves of the process heating exchangers, such as steam-heater, and reboiler. The left side of TSP shows the composite curves of the process cooling exchangers, such as steam-generator, cooler and condenser.

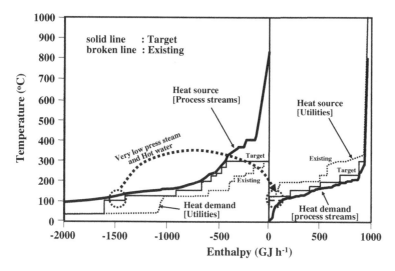

Fig. 6.1 Total site profile analysis for selected block in Mizushima industrial area

6.2.2 R-curve Analysis

Makwana et al. (1998) developed Top-level analysis that could be applied for existing total site utility systems in terms of current performance and the potential scope for improvement. Subsequently, a method for analyzing and optimizing energy systems was developed by Kimura (1998). This method builds on the concepts of R-curve by Kenney (1984) and Top-level analysis by Makwana et al. (1998). The R-curve analysis method was further developed by Kimura and Zhu (2000) to determine the most economical modifications to existing utility systems. The R-curve provides a target for the efficiency of utility system by converting fuel energy into heat (Q_{heat}) and power (W). The integrated energy efficiency (Eq. 6.1), which is the fuel utilization efficiency, is defined as a ratio of the useful part of energy and the integrated energy consumption (Q_{fuel}). The shape of the R-curve is determined by the fact that the production of shaft power from fuel energy requires a heat sink and, in an integrated site, the process plant acts as the heat sink for power generation. The larger the heat demand relative to power demand, the more efficient the overall generation becomes. This is represented by the R-ratio—the ratio of power to heat demand from the process (Eq. 6.2) under the operating condition of the site.

$$\text{Integrated energy efficiency} = (W + Q_{heat})/Q_{fuel} \qquad (6.1)$$

$$\text{R-ratio (power-to-heat ratio)} = W/Q_{heat} \qquad (6.2)$$

Figure 6.2 shows the theoretical limit lines for two energy systems. One is the "Gas turbine combined system" and the other is the "Boiler and turbine conventional system". Figure 6.3 illustrates graphically the definition of the two key parameters.

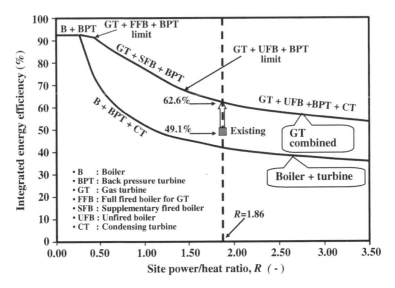

Fig. 6.2 R-curve analysis for selected block in Mizushima industrial area

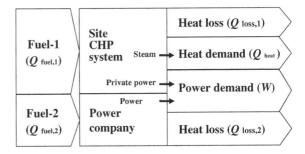

Fig. 6.3 Two key parameters in R-curve analysis

For a given site R-ratio, the R-curve shows the maximum achievable efficiency. The difference between the existing efficiency and maximum efficiency shows the potential of improvement. R-curves can be built up for individual site and the power and heat demands of multiple sites can be combined to determine complex wide opportunities. The application of pinch technology to save thermal energy consumption will interact with the R-curve analysis, as reduced steam demand will increase the R-ratio.

In the conventional approach, R-curve analysis is applied on the condition of the defined minimum energy requirement balance to determine the ultimate energy saving potential in the area-wide integration.

To perform a fully rigorous complex-wide assessment requires process optimization of each individual plant using pinch technology analysis, with the resultant TSP profiles being used to optimize the utility system and provide the basis for the R-curve analysis. This would require a very large amount of data collection and pinch analysis work.

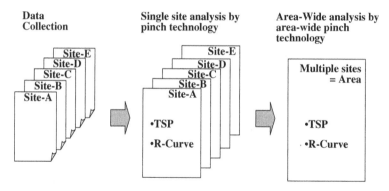

Fig. 6.4   Area-wide analysis procedure

In this study, the more practical "Grey Box Approach" (Brown 1999) was utilized. Many efforts have already been applied to the individual sites to improve energy efficiency and the sites are believed to be the most efficient in the world. Therefore, modification of heat recovery within the individual processes was not considered. Instead, the heat exchangers of the Process-Utility interfaces (e.g., heaters, coolers and steam generators) were used to generate the TSP curves. These curves were used as an aid to develop ideas for a practicable energy saving project.

6.2.3 Work Procedure

Figure 6.4 shows the procedure for applying area-wide pinch technology to an industrial area composed of multiple sites. The operation data for the TSP analysis and the R-curve analysis are collected initially from all the sites in the industrial area. The data are then analyzed by using the TSP analysis and the R-curve analysis with respect to each single site. The results of the analyses determine which site has the highest performance and which site has the lowest performance from the point of view of energy saving. The data are then combined and condensed into one virtual site and then again analyzed by the TSP and the R-curve analyses, which eventually evaluate the total energy saving potential in an industrial area composed of multiple sites.

6.3 Application of Area-Wide Pinch Technology

"Area-wide pinch technology", which consists of TSP analysis and R-curve analysis, was applied to two of the biggest heavy chemical complexes in Japan, i.e., Mizushima and Chiba industrial areas.

6.3.1 Mizushima and Chiba Industrial Areas

Mizushima industrial area is located in west part of Japan (150 km west from Osaka, the second largest city in Japan). It has 35 individual sites consisting of refineries, petrochemical plants, chemicals, and others. The area is divided into block A (15 sites), block B (10 sites) and block C (10 sites). Each block is separated by canals.

Chiba industrial area is located 30 km east from Tokyo across Tokyo bay. In this area, the cooperation was obtained from a total of 23 sites consisting of process industries including petrochemical, refinery and power companies. The area is also divided into block A (five sites), block B (six sites) and block C (12 sites), in the same way as Mizushima industrial area.

6.3.2 Results

6.3.2.1 Total Site Profile Analysis

Figure 6.1 shows the result of TSP analysis for the selected block in Mizushima industrial area. It was found that unutilized exhaust heat exists in the range between 100 and 120 °C. A combination of very LPS (low-pressure steam) and hot water can be recovered as shown in left side of Fig. 6.1. Recovering this steam and hot water will reduce utility consumption of the LPS from the utility plant.

The energy saving potential by TSP analysis depends upon how to install the new utility conditions of the heat demand and the heat supply to make the composite curves come closer each other. The conventional TSP analysis for single site uses the existing utility conditions and sometimes adds a few of new utility condition. However, the area-wide TSP analysis for multiple sites uses the greatest common divisor of all the existing utility conditions in the corresponding sites because the energy saving potential should be counted larger. It especially adds some of the low-temperature utility conditions, such as very LPS and hot water because unutilized exhaust heat should be utilized in multiple sites for further energy saving.

Figure 6.1 shows a large gap between the composite curves. By approaching the header conditions from the existing line to the targeting lines in the left side of TSP chart, the higher temperature steam would be recovered and produce further power generation because the suction steam pressure for the steam turbine generator is increased and the thermal pressure drop is increased. A large gap in the right side of the TSP chart would also lead to a new power generation opportunity. These energy saving potentials (new power generation opportunities) have not been included in the total area-wide potential because they do not result in less energy consumption, only more effective conversion of fuel into power.

Table 6.1 Results of energy saving studies

Industrial area	Mizushima	Chiba	Mizushima /Chiba
1 No. of sites	35	23	1.5
2 Integrated fuel consumption (fuel + power)	3,785,000 kL y^{-1}	2,880,000 kL y^{-1}	1.3
3 Energy saving potential by R-curve analysis	1,002,000 kL y^{-1}	508,000 kL y^{-1}	2.0
4 Energy saving potential by TSP analysis	210,700 kL y^{-1}	133,000 kL y^{-1}	1.6
Total	1,212,700 kL y^{-1}	641,000 kL y^{-1}	
Domestic crude oil consumption	2 days	1 day	

kL y^{-1} : Kiloliter per year: conversion to annual crude oil
1 kL = 38.8 GJ
Domestic crude oil consumption in Japan: 600,000 kL day-1 (approximately)

6.3.2.2 R-curve Analysis

The utility system is currently based on boilers and steam turbines, with a small size of gas turbine. All heat and power are provided by private utility plants with power generation systems. It can be seen in Fig. 6.2 that the utilities efficiency is just over the theoretical curve for a B (boiler) + BPT (back pressure turbine) + CT (condensing turbine) system and is located at the region of GT (gas turbine) + UFB (unfired boiler) + BPT + CT. This suggests that the generation and use of steam is not so efficient because a large power demand reduces the efficiency due to the loss of condensing power generation. However, the lack of adequate size of gas turbines in the utility plant results in an efficiency gap as the existing point is substantially below the upper line that can be achieved in ideal gas turbine (GT) combined systems.

Figure 6.2 shows the result of R-curve analysis for existing utility consumption of the selected block of Mizushima industrial area. The introduction of the ideal "Gas turbine combined system" could increase the integrated energy efficiency from 49.1 to 62.6 %. The increment means as well that the fuel consumption would be decreased by 13.5 % while maintaining the present heat and power demands.

6.3.2.3 Comparison of Mizushima and Chiba Industrial Areas

Table 6.1 shows the summary of Mizushima industrial area and Chiba industrial areas. Mizushima has 35 sites and its integrated fuel consumption, which is the sum of the fuel consumption and the electrical power consumption in the energy system, is 3,785,000 kL y^{-1} (146.9 × 10^6 GJ y^{-1}). The R-curve analysis determined that Mizushima has potentially 1,002,000 kL y^{-1} (38.9 × 10^6 GJ y^{-1}) of energy saving, while the TSP analysis shows that this area has a potential

210,700 kL y^{-1} (8.2 × 10^6 GJ y^{-1}) of energy saving. Mizushima therefore has the potential to save a total of 1,212,700 kL y^{-1} (47.1 × 10^6 GJ y^{-1}) of energy, which is equivalent to almost 2 days of crude oil consumption in Japan.

Table 6.1 also shows the result for Chiba. The energy saving potential in Chiba area is equivalent to almost 1 day of crude oil consumption in Japan. In comparison, Mizushima industrial area has 1.5 times more sites than Chiba industrial area but its integrated fuel consumption is only 1.3 times larger than Chiba. In contrast, the energy saving potential of Mizushima is almost twice that of Chiba. The result made it apparent that the equipment in the energy system in Mizushima industrial area performs less efficiently than that in Chiba. This is supported by the fact that Mizushima industrial area was established much earlier than Chiba industrial area.

6.4 Area-Wide Integration Projects for Energy Saving

A number of area-wide integration project ideas for energy saving have been identified from the result of above area-wide pinch technology study. Those ideas were developed under the concepts of heat utilization in two temperature zones; higher and lower temperature. In the higher temperature heat zone, the energy system was integrated and the higher temperature steam was shared, which was named as the "energy sharing system".

In lower temperature heat zone, the surplus heat recovered from one site could be shared with adjacent sites, which was named as the "heat sharing system". After utilization by the heat sharing system, there was a large amount of low-grade heat (<150 °C) which was considered as waste heat. Low heat power generation system was developed utilizing such low-grade heat.

6.4.1 Energy Sharing System

Figure 6.5 shows energy sharing system in sites A and B. Site A is a refinery and has four boilers that generate high-pressure steam. Heavy fuel oil was used to power those boilers. The issue in site A was the No.4 boiler (B-4) that had lower efficiency than the others and its annual maintenance cost was expensive.

Site B, which is adjacent to site A, is a petrochemical plant and a large amount of high-pressure steam was recovered by steam generators in process systems. The recovered steam was supplied to condensing turbines to produce the shaft power for rotating machines. For example, No.5 condensing turbine (T-5) supplied its shaft power to a compressor. The issue in site B was not being able to develop new idea to improve the energy efficiency. The issues of two of the sites were identified during the area-wide pinch technology study. To solve their issues, an energy sharing system project between sites A and B was developed.

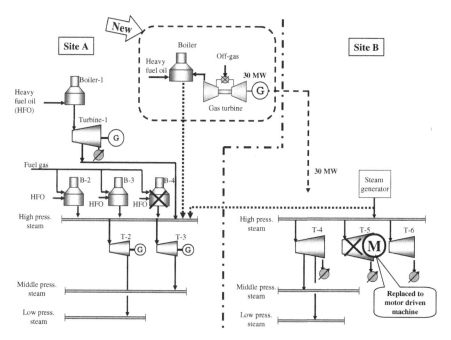

Fig. 6.5 Energy sharing system

Firstly, because a condensing turbine has a lower efficiency in thermodynamics, around 10–15 %, the compressor driver operated by T-5 was disconnected and changed to a motor-driven type, which led to the energy efficiency of the compressor being increased to approximately 90 %. However, the recovered high-pressure steam that was used for T-5 would then be surplus. As it had been planned that B-4 should be idle in operation due to its lower efficiency and expensive maintenance cost, the high-pressure steam could be supplied to a steam header in site A. Unfortunately, the amount of the high-pressure steam supplied from site B was not sufficient for the amount that B-4 had used to generate, and the new motor driven compressor needed a certain amount of electric power. In order to solve such a new heat and power requirement, a new facility was installed which is shown surrounded by a dotted line in Fig. 6.5. A new CHP facility was installed which had a gas turbine (GT) system equipped with a full fired boiler (FFB). The off-gases from both sites were collected and supplied to the GT as a fuel and the heavy fuel oil supply for B-4 was rerouted to the FFB. As the flue gas from GT system was sent to the FFB, utilizing not only its heat but also its residual oxygen for combustion air, the total energy efficiency of the CHP facility, with a combined system of GT and FFB, was able to become very high. The high-pressure steam from the FFB was sent to site A and was able to compensate the steam balance in Site-A. The GT system supplied the electric power to the compressor that used to be driven by T-5. It had therefore become unnecessary to purchase expensive power from the power company in site B.

Eventually, installing the new CHP facility consisting of GT and FFB was able to achieve a large amount of energy saving. It was expected in case of Fig. 6.5 that the amount of energy saving would be 30,600 kL y^{-1} (1.2 × 10^6 GJ y^{-1}). Both sites A and B invested in this new CHP facility.

6.4.2 Heat Sharing System

Figure 6.6 shows heat sharing system in sites C and D, which were under separate operation. The result of TSP analysis confirmed that there was an energy-saving potential by the heat-sharing system. Site C used to have an air-fin cooler to cool the hot process stream and discarded the heat to atmosphere as waste heat. It was found that hot water (90–95 °C) was able to be recovered from the hot process stream. Site D had a boiler and utilized several pressure levels of steam. Low-pressure steam was supplied to a steam heater in which the cold process stream was heated. It was found that the cold process stream was able to be heated sufficiently by the hot water because the required temperature of the cold process stream in the steam heater was lower than temperature of hot water.

It was planned to recover hot water in site C and supply it to the existing steam heater in site D and therefore, the low-pressure steam was replaced by the hot water from site C. Eventually consumption of some of the low pressure steam was able to be reduced in site D. This heat sharing system achieved 3,400 kL y^{-1} (0.13 × 10^6 GJ y^{-1}) of energy saving. As can be seen in Fig. 6.6, sites C and D had already completed energy saving measures and it was considered that there was no way to make use of the low-grade heat. However, it would have been able to have a new energy saving project by introducing the concept of area-wide heat sharing. The amount of energy saving achieved from this project was shared between both sites C and D.

6.4.3 Low Heat Power Generation System

When the utilization of low-grade heat (lower than 150 °C) is being considered, it would be advisable to first check by using total site profile analysis if there is a neighboring user for such heat. If such a user is found, a heat sharing system shall be developed. If no neighboring user for such heat is found, it would be possible to install a low heat power generation system to convert such low-grade heat into electric power.

Figure 6.7 shows the process flow sheet of low heat power generation system. This site has a fractionator and its overhead vapor (around 120 °C) used to be cooled by air fin coolers. The heat duty of the air fin coolers was very large and was not utilized but discharged to atmosphere. A low heat power generation system was considered with the expectation of 8–10 % in power generation efficiency.

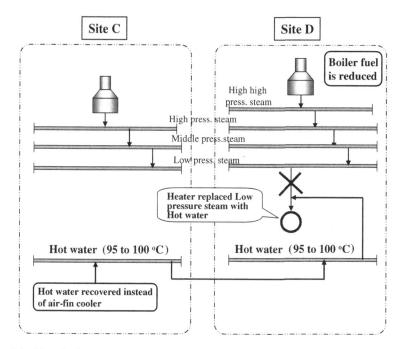

Fig. 6.6 Heat sharing system

In Fig. 6.7, the overhead vapor from the fractionator was introduced into vaporizer in which it exchanged with the high concentration of ammonia solution (more than 90 wt%) as the working fluid. At the vaporizer, the high pressure ammonia vapor, 3 MPag approximately, was generated and the cooled overhead vapor was returned to the original overhead cooling system. The high pressure ammonia vapor was introduced to a separator to remove the liquid and then introduced into a turbine generator to produce the electric power. Before it was introduced into the turbine, the ammonia vapor was superheated by low-pressure steam to avoid the erosion in the turbine. The exhausted vapor from the turbine was mixed with the separator liquid that was cooled in a regenerator, and then cooled down further by a condenser. In the condenser the working fluid was converted fully into liquid. The working fluid was sent to vaporizer again through the regenerator by a pump and then came back around this closed system.

6.5 Discussion

Total site profile (TSP) analysis and R-curve analysis of area-wide pinch technology are used for area-wide energy integration. The R-curve analysis is used to analyze the utility system and to confirm the energy efficiency under the present

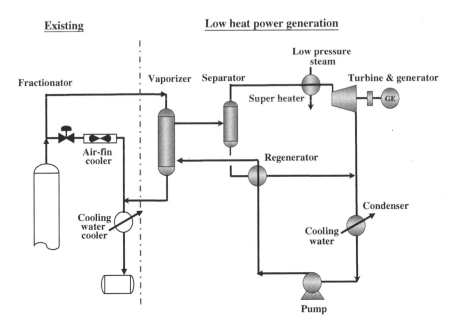

Fig. 6.7 Low heat power generation system

condition and the R-curve analysis suggests the more efficient use of energy. The TSP analysis is used to evaluate the present situation of heat utilization and to identify the opportunity of heat sharing projects. As shown in Table 6.1, area-wide pinch technology is able to conduct macro evaluation on multiple sites.

As shown in Fig. 6.5, energy-sharing system leads to a large energy saving, such as 30,600 kL y^{-1} (1.2 × 10^6 GJ y^{-1}). The project cost is huge because new gas turbines and boilers are expensive, but the profit is large. Despite the good economy and efficiency, there are some obstacles to overcome in the implementation of a new large utility facility, such as how to invest, who and how to operate, and how to calculate and allocate the benefits.

Conversely, the heat sharing system shown in Fig. 6.6, which consists of heat exchangers and piping, does not cost so much but provides good economically efficient utilization of low-grade heat. It would have few influences on the process system.

It is expected, based on the result of area-wide pinch technology, that heat sharing system projects would first be implemented in adjacent sites and then an energy sharing system would follow on later. After utilizing the low-grade heat in adjacent sites, the heat still remaining in such low-grade heat would be converted to electric power by using a low heat power generation system.

References

Bandyopadhyay S, Varghese J, Bansal V (2010) Targeting for cogeneration potential through total site integration. Appl Therm Eng 30(1):6–14

Brown SM (1999) The drive for refinery energy efficiency. Petroluem Technology Quaterly Refining Autumn 45–55

Dhole VR, Linnhoff B (1992) Total site targets for fuel, co-generation, emissions, and cooling. Comput Chem Eng 7(Suppl.1):s101–s109

Kenney WF (1984) Energy conservation in the process industry. Academic Press, Orlando

Kimura H (1998) R-curve concepts for analysis and optimisation of cogeneration systems. MSc Dissertation, UMIST, UK

Kimura H, Zhu XX (2000) R-curve concept and its application for industrial energy management. Ind Eng Chem Res 39(7):2315–2335

Klemes J, Dhole VR, Raissi K, Perry SJ, Puigjaner L (1997) Targeting and design methodology for reduction of fuel, power and CO2 on total sites. Appl Therm Eng 7:993–1003

Makwana Y, Smith R, Zhu X.X (1998) A novel approach for retrofit and operation management of existing total sites. Comput Chem Eng 22:s793–s796

Matsuda k (2008) The development of energy sharing in industrial areas of Japan with pinch technology. J Chem Eng Japan 41(10):992–996

Perry S, Klemes J, Bulatov I (2008) Integrating waste and renewable energy to reduce the carbon footprint of locally integrated energy sectors. Energy 33(10):1489–1497

Raissi K (1994) Total site integration. Ph.D. Thesis, UMIST, UK